ENERGY SCIENCE, ENGINEERING AND TECHNOLOGY

GUIDELINES FOR IMPROVED DUCT DESIGN AND HVAC SYSTEMS IN THE HOME

Energy Science, Engineering and Technology

Additional books in this series can be found on Nova's website
under the Series tab.

Additional E-books in this series can be found on Nova's website
under the E-book tab.

ENERGY SCIENCE, ENGINEERING AND TECHNOLOGY

GUIDELINES FOR IMPROVED DUCT DESIGN AND HVAC SYSTEMS IN THE HOME

LANE M. COLLINS
AND
DANIELLE E. MARTINEZ
EDITORS

Nova Science Publishers, Inc.
New York

For permission to use material from this book please contact us:
Telephone 631-231-7269; Fax 631-231-8175
Web Site: http://www.novapublishers.com

Additional color graphics may be available in the e-book version of this book.

Library of Congress Cataloging-in-Publication Data

ISBN: 978-1-62081-443-7

Published by Nova Science Publishers, Inc. † New York

CONTENTS

PREFACE

Heating and cooling losses from forced-air ducts can result in high energy costs, lead to thermal comfort problems, and–in some extreme situations–result in serious health and safety concerns. Reducing air leakage and conductive losses from ducts can be a straight-forward way to reduce energy use and improve comfort in homes. This book discusses the basics of air distribution and duct design; strategies to seal and insulate ducts in existing homes; accurate heating and cooling load calculations and HVAC equipment sizing.

Chapter 1 - Right-sizing of a heating, ventilation, and air-conditioning (HVAC) system is the selection of equipment and the design of the air distribution system to meet the accurate predicted heating and cooling loads of the house. The estimated heating and cooling loads are those required to meet the inside design conditions on the design load day. The design load day is not the most extreme weather conditions or the conditions that represent the majority of hours in a year. Temperature swings are expected in the conditioned space during extreme weather, and the system must be able to deliver comfort during the many hours of partial load conditions.

The higher performing house with higher levels of insulation, more energy efficient windows, reduced air infiltration, and controlled mechanical ventilation has a lower load, and consequently less air volume is needed to condition the space. This presents both challenges and opportunities when designing the air distribution system. Challenges arise when using air outlets sized by "rule of thumb" that do not have the throw needed to provide air mixing in the room to achieve the desired comfort results. At the same time, the energy efficient house design process offers opportunities to integrate the HVAC system into the design earlier in the process and to utilize a strategy of shorter duct runs with smaller cross-sectional areas that can reduce system costs and that can be installed completely within the conditioned space.

To illustrate the air distribution basics and the issues faced when implementing a robust duct design methodology for an energy efficient house, two theoretical houses that meet the 2009 International Energy Conservation Code (IECC) prescriptive path were modeled. The first was modeled in the IECC Climate Zone 5 - CZ5, Chicago, Illinois. The second was modeled in IECC Climate Zone 2 - CZ2, Orlando, Florida.

The performance criteria of a well-designed air distribution system include air delivery without inducing drafts or excessive noise. The occupied zone of a room, 2 feet away from walls and 2 feet off the ceiling, is the area where the perception of a draft must be avoided. Noise in the air distribution system comes from the velocity of air in the ducts, supply, or return and the air handling equipment itself. Strategies to avoid the comfort issues perceived as drafts or system noise are critical to the design of the HVAC system.

The air distribution system will give consideration to the return side of the system as well as the supply side. A well-designed return air strategy is critical for the performance of the HVAC system. Rooms without adequate return air can impede supply airflow due to overpressurization in the room, leading to comfort issues. Every room with a supply outlet (with the exception of bathrooms or kitchens due to the potential for spreading odors through the house) must have a clear return air pathway.

The selection and placement of the supply air outlets are critical to the comfort in the space. With the lower loads of the energy efficient house, the volume of air, measured in cubic feet per minute, required to condition the space is also lower, presenting challenges in selecting the air supply outlets. The air must be delivered in a manner that mixes the supply air with the room air without causing the sensation of a draft on the occupant.

Finally, decisions made during the early design phase are critical to the successful performance of the HVAC system. Early consideration of the HVAC system in an integrated design allows the designer to allocate adequate space for the equipment and ducts while identifying principal potential conflicts between the building's structure and the HVAC system.

Chapter 2 - Ducts carry conditioned air between a central air handler and rooms throughout a home. In an ideal system, all of the supply air leaving the air handler would be transferred by the duct system to various points of use in the home and ultimately be discharged through supply registers within the rooms. Similarly, all of the return air in an ideal system would be drawn from the appropriate return air grilles, through return ducts, and back to the furnace or air handler. Historically, this has not been the case. According to the EPA ENERGY STAR® program, the typical home has 20% duct leakage. In older homes, it is not uncommon for duct leakage to account for 30%-50% of the total system air flow. Much higher leakage rates are possible in very poorly sealed systems. Ideally, duct leakage would be negligible, but a goal of less than 5% of the total system air flow is reasonable for newly installed systems. For existing duct systems, when parts of the duct system are not accessible, going from 30% leakage to 5% leakage is often not realistic. Ducts should be sealed as tightly as is practical, and this level varies tremendously from system to system.

Chapter 3 - The *Strategy Guideline: Accurate Heating and Cooling Load Calculations* report provides information for the following groups:

- Heating Ventilation and Air Conditioning (HVAC) Mechanical Contractors
- HVAC System Designers
- Builders
- House Remodelers.

This guide can be used as a point of entry for collaborative discussions between a builder, HVAC contractor, and other trade partners to understand the importance of appropriately calculating peak heating and cooling loads, as the first step of HVAC system design.

Accurate load calculations have a direct impact on energy efficiency, occupant comfort, indoor air quality, and building durability. The load calculation is the first step of the iterative HVAC design procedure, as a full HVAC design involves much more than just the load calculation. The loads modeled by the heating and cooling load calculation process will dictate the equipment selection and duct design to deliver conditioned air to the rooms of the house. This guide references the methodologies of the Air Conditioning Contractors of America (ACCA) publication *Manual J Residential Load Calculation Eighth Edition* (ACCA MJ8), which in turn references information provided by the American Society of Heating, Refrigerating and Air-Conditioning Engineers (ASHRAE). ACCA MJ8 only applies to single family detached dwellings, low-rise condominiums, and townhouses.

This guide is not a new method for performing load calculations, a substitute for established methods of performing load calculations, or step-by-step instructions on how to perform the load calculations. This guide presents the key criteria required to create accurate heating and cooling load calculations and offers examples of the implications when inaccurate adjustments are applied to the HVAC design process. The guide shows, through realistic examples, how various defaults and arbitrary safety factors can lead to significant increases in the load estimate. Emphasis is placed on the risks incurred from inaccurate adjustments or ignoring critical inputs of the load calculation.

Chapter 4 - Right-sizing of a residential heating, ventilation, and air conditioning (HVAC) system involves the selection of equipment and the design of the air distribution system to meet the accurate predicted heating and cooling loads of a house. This Strategy Guideline follows the Air Conditioning Contractors of America (ACCA) *Manual S—Residential Equipment Selection* (Manual S) (Rutkowski 1995) to describe what information is needed and how to use that information to initially select the equipment for a properly designed HVAC system. Equipment selection and duct design are iterative processes where the capacity of the equipment is balanced against the design of the distribution system.

This Strategy Guide describes the equipment selection of a split system air conditioner and furnace for an example house in Chicago, Illinois, as well as a heat pump system for an example house in Orlando, Florida. The required heating and cooling load information for the two example houses was developed by Burdick (2011).

Selection of the equipment can have a substantial impact on the efficiency and operating costs of the system. For the example houses presented in this Strategy Guide, a 92.5% AFUE (annual fuel utilization efficiency) furnace with a 16 SEER (seasonal energy efficiency ratio) air conditioner was the system chosen for the Chicago House and a 9.2 HSPF (heating seasonal performance factor)/16 SEER heat pump for the Orlando House.

With an accurate estimate of the heating and cooling loads completed, the HVAC system designer can begin the equipment selection procedure. It is the system designer's responsibility to select the heating and cooling equipment and to verify that the selected pieces of equipment meet the following:

- The total cooling load, composed of both the sensible load (temperature) and the latent load (humidity)
- The heating load
- The blower capacity (in cubic feet per minute [cfm]) to meet the volume of air cfm range needed.

The general steps to equipment selection are:

1. Determine accurate heating and cooling loads.
2. Select cooling equipment.
3. Select heating equipment.
4. Design the ducts.

It is worth noting that the design of a space conditioning system is an iterative process. Once a duct system configuration and preliminary design are completed, different equipment may need to be selected to achieve the proper airflow to satisfy the room-by-room heating and cooling loads.

In: Guidelines for Improved Duct Design … ISBN 978-1-62081-443-7
Editors: L. M. Collins, D. E. Martinez

Chapter 1

ADVANCED STRATEGY GUIDELINE: AIR DISTRIBUTION BASICS AND DUCT DESIGN*

Arlan Burdick

DEFINITIONS

ACCA	Air Conditioning Contractors of America
ANSI	American National Standards Institute
ASHRAE	American Society of Heating, Refrigerating and Air-Conditioning Engineers
Available Pressure	The difference between the static pressure and the pressure against which the selected equipment can deliver the cubic feet per minute of air. This value is what is available in the design for duct and supply outlet pressure losses.
Btu	British thermal unit
Btu/h	Btu per hour
CF	Cooling factor
cfm	Cubic feet per minute

* This is an edited, reformatted and augmented version of Office of Energy Efficiency and Renewable Energy, U.S. Department of Energy, dated December 2011.

Coanda Effect	The tendency of moving air to attach itself to a surface and flow along it. As air from the supply outlet moves across the ceiling surface, a certain amount of friction occurs between the supply air and the ceiling surface. This resistance to the flow of the supply air pulls the supply air toward the ceiling surface, causing it stick to the surface.
Effective Length	The length of duct used for design purposes that includes the straight length of duct added to all the fitting equivalent lengths.
Equivalent Length	The amount of duct length added to a duct run by fittings. Often the equivalent length of a fitting can be equal to or longer than the straight duct lengths due to the additional friction of turning air.
Face Velocity	The velocity of supply air as it leaves the supply outlet.
Friction Rate	The pressure loss between two points in a duct system that are separated by a specific length. The units for friction rate are inches water column per 100 feet (iwc/100 ft).
ft/min	Feet per minute
HF	Heating factor
HVAC	Heating, ventilation, and air-conditioning
IECC	International Energy Conservation Code
iwc	Inch water column
Manual D	*ACCA Manual D—Residential Duct Systems*
Manual J	*ACCA Manual J—Residential Load Calculation*
Manual S	*ACCA Manual S—Residential Equipment Selection*
Manual T	*ACCA Manual T—Air Distribution Basics for Residential and Small Commercial Buildings*
NC	Noise criteria
Spread	The width of the air stream leaving the supply outlet as determined by the characteristics of the supply outlet.
Static Pressure	The amount of pressure built up in the system by accessories such as the cooling coil or filters.
Terminal Velocity	The velocity of supply air at the end of its effective throw.

Throw — The effective distance that air leaving the supply outlet can reach.

Executive Summary

Right-sizing of a heating, ventilation, and air-conditioning (HVAC) system is the selection of equipment and the design of the air distribution system to meet the accurate predicted heating and cooling loads of the house. The estimated heating and cooling loads are those required to meet the inside design conditions on the design load day. The design load day is not the most extreme weather conditions or the conditions that represent the majority of hours in a year. Temperature swings are expected in the conditioned space during extreme weather, and the system must be able to deliver comfort during the many hours of partial load conditions.

The higher performing house with higher levels of insulation, more energy efficient windows, reduced air infiltration, and controlled mechanical ventilation has a lower load, and consequently less air volume is needed to condition the space. This presents both challenges and opportunities when designing the air distribution system. Challenges arise when using air outlets sized by "rule of thumb" that do not have the throw needed to provide air mixing in the room to achieve the desired comfort results. At the same time, the energy efficient house design process offers opportunities to integrate the HVAC system into the design earlier in the process and to utilize a strategy of shorter duct runs with smaller cross-sectional areas that can reduce system costs and that can be installed completely within the conditioned space.

To illustrate the air distribution basics and the issues faced when implementing a robust duct design methodology for an energy efficient house, two theoretical houses that meet the 2009 International Energy Conservation Code (IECC) prescriptive path were modeled. The first was modeled in the IECC Climate Zone 5 - CZ5, Chicago, Illinois. The second was modeled in IECC Climate Zone 2 - CZ2, Orlando, Florida.

The performance criteria of a well-designed air distribution system include air delivery without inducing drafts or excessive noise. The occupied zone of a room, 2 feet away from walls and 2 feet off the ceiling, is the area where the perception of a draft must be avoided. Noise in the air distribution system comes from the velocity of air in the ducts, supply, or return and the air handling equipment itself. Strategies to avoid the comfort issues perceived as drafts or system noise are critical to the design of the HVAC system.

The air distribution system will give consideration to the return side of the system as well as the supply side. A well-designed return air strategy is critical for the performance of the HVAC system. Rooms without adequate return air can impede supply airflow due to overpressurization in the room, leading to comfort issues. Every room with a supply outlet (with the exception of bathrooms or kitchens due to the potential for spreading odors through the house) must have a clear return air pathway.

The selection and placement of the supply air outlets are critical to the comfort in the space. With the lower loads of the energy efficient house, the volume of air, measured in cubic feet per minute, required to condition the space is also lower, presenting challenges in selecting the air supply outlets. The air must be delivered in a manner that mixes the supply air with the room air without causing the sensation of a draft on the occupant.

Finally, decisions made during the early design phase are critical to the successful performance of the HVAC system. Early consideration of the HVAC system in an integrated design allows the designer to allocate adequate space for the equipment and ducts while identifying principal potential conflicts between the building's structure and the HVAC system.

1. Overview

The heating, ventilation, and air-conditioning (HVAC) system is arguably the most complex system installed in a house and is responsible for a substantial component of the total house energy use. A right-sized HVAC system will provide the desired comfort and will run efficiently. Right-sizing of an HVAC system is the selection of equipment and the design of the air distribution system to meet the accurate predicted heating and cooling loads of the house. Right-sizing the HVAC system begins with an accurate understanding of the heating and cooling loads on a space; however, a full HVAC design involves more than just the load estimate calculation; the load calculation is the first step of the iterative HVAC design procedure. Heating and cooling loads are dependent upon the building location, sighting, and the construction of the house, whereas the equipment selection and the air distribution design are dependent upon the loads and each other.

Is not a procedural step-by-step set of instructions on how to create a duct design; rather, it The procedures of residential HVAC design (see Figure 1) are covered in detail by a series of publications produced by the Air Conditioning Contractors of America (ACCA), which in turn references information provided by the American Society of Heating, Refrigerating and Air-Conditioning Engineers (ASHRAE). The heat loss and gain values are estimated using the procedures from the ACCA publication, *Manual J—Residential Load Calculation*, Eighth Edition (Manual J) (Rutkowski 2006). Manual J applies only to single-family detached dwellings, low-rise condominiums, and town houses. Mechanical equipment selection is done with the aid of the ACCA *Manual S—Residential Equipment Selection* (Manual S) (Rutkowski 1995). The ductwork to convey the proper amount of conditioned air to meet the load requirements of the space can be designed with the aid of the ACCA *Manual D—Residential Duct Systems* (Manual D) (Rutkowski 2009). Conditioned air delivery to the space is controlled by the type and size of the air outlet, as discussed in the ACCA *Manual T—Air Distribution Basics for Residential and Small Commercial Buildings* (Manual T) (Rutkowski 1992). Manual T provides guidance on selecting the air outlet size, type, and location. The ACCA procedures have been written into commercial software packages to help the designer work through the iterations required for a good design. While commercial software is an important tool for design, it should be operated with a solid understanding of the procedures and intentions of proper HVAC design.

For the purposes of this Strategy Guideline, an energy efficient house is defined as one that is designed and built for decreased energy use and improved comfort through higher levels of insulation, more energy efficient windows, high efficiency space conditioning and water heating equipment, energy efficient lighting and appliances, reduced air infiltration, and controlled mechanical ventilation. Specification levels for energy efficient houses have historically been prescribed by beyond code programs that set a percentage better than code for energy use, such as ENERGY STAR®, which requires houses to be 15% more energy efficient than code. Beyond code programs continue to set a percentage better than the improved codes for energy use, raising the bar for whole house energy efficiency. The 2009 International Energy Conservation Code (IECC) establishes an estimated 15% improvement in energy efficiency over the previous 2006 IECC requirements. Ongoing code cycles incrementally increase the minimum efficiency of a house. For example, the 2012 IECC achieves approximately 30% savings over the 2006 version. As the new codes are adopted and implemented, a house that was built under an above code program in 2010 will likely be the code mandated house in 2015. As the energy efficiency of the house is increased under code or above code programs, the peak heating and cooling loads are significantly reduced.

To illustrate the procedures of HVAC equipment selection, two theoretical houses that meet the 2009 IECC prescriptive path were modeled. The first house (the "Chicago House") is a one-story, 2,223-square-foot (above-grade) house that has a full conditioned basement and is located in Chicago, Illinois (IECC Climate Zone 5 - CZ5). The other house (the "Orlando House") is a one-story 2,223-square-foot slab-on-grade house in Orlando, Florida (IECC Climate Zone 2 - CZ2).

In addition to the climatic differences between the two house locations, the Chicago House is a house over a full basement, whereas the Orlando house is a slab-on-grade foundation, allowing for the discussion of different mechanical and ductwork layouts.

The impacts and importance of accurately estimating heating and cooling loads are covered in depth in the U.S. Department of Energy Building America *Strategy Guideline: Accurate Heating and Cooling Load Calculations* (Burdick 2011a). The equipment selection to satisfy those estimated loads is discussed in the U.S. Department of Energy Building America *Strategy Guideline: HVAC Equipment Sizing* (Burdick 2011b). This Strategy Guideline on air distribution basics and duct design for an energy efficient house builds upon the information developed in those two previous guides. Results of the load calculations and the equipment selected in the two previous guidelines are used for the air distribution examples here.

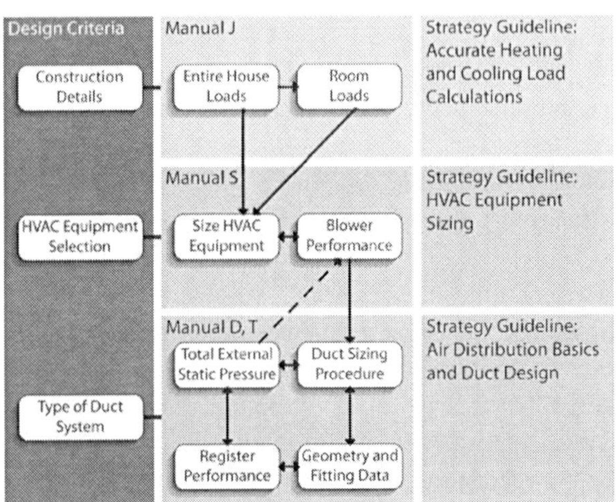

Figure 1. HVAC design flow.

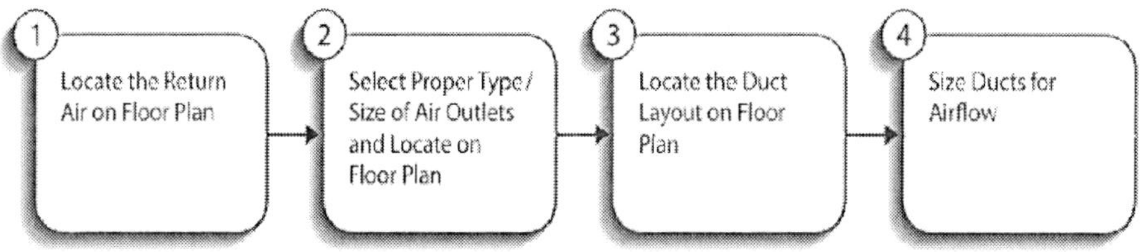

Figure 2. General steps for air distribution design.

The estimated heating and cooling loads are those required to meet the inside design conditions on the design load day. The design load day is not the most extreme weather condition or the conditions that represent the majority of hours in a year. Therefore, temperature swings are expected in the conditioned space during extreme weather, and the system must be able to deliver comfort during the many hours of partial load conditions. The higher performing house has a lower load, and consequently less air volume is needed to condition the space. This presents both challenges and opportunities when designing the air distribution system. Challenges arise when using air outlets sized by "rule of thumb" that do not have the throw needed to provide air mixing in the room to achieve the desired comfort results. At the same time, the energy efficient house design process offers opportunities to integrate the HVAC system into the design earlier in the process and to utilize a strategy of shorter duct runs with smaller cross-sectional areas that can reduce system costs and that can be more easily installed completely within the conditioned space.

In general, the primary considerations for the air distribution design are shown in Figure 2.

2. Performance Criteria

Occupant comfort is the overall objective of the HVAC system, and air distribution is the means of delivering conditioned air to provide comfort in the room. Comfort is very subjective and can vary by individual activities and metabolic rates. The principal comfort criteria related to the air distribution system are temperature mixing and uniformity, drafts that can be created by the supply outlets, and system noise.

2.1. Temperature Mixing and Uniformity

Temperature in the room is maintained by delivering just the proper amount of air in a manner in which this air can mix with the air that is already in the room. ACCA (2009) offers guidance for dry-bulb temperature variances from the thermostat setting during the heating season as measured at the thermostat to be ±2°F. Similarly the temperature during the heating season in any room should be ±2°F of the thermostat set temperature. Room-to-room temperature differences or floor-to-floor temperature differences should be no greater than 4°F in the heating season. ACCA guidance for room temperature variances from the thermostat setting is ±3°F during the cooling season.

2.1.1. Volume and Pressure of Air

The amount of air available to provide comfort to the rooms is determined by the equipment selected to meet the loads of the house. In energy efficient houses with lower loads, properly selected equipment will have less capacity to supply air, and the performance of the system is dependent upon making the best use of the available air. The volume of air the equipment is rated to deliver is determined at a specific pressure. Every item, including the ductwork used in the airway, adds to the pressure loss within the system.

2.1.2. Ductwork Materials and Configuration

The ductwork that is used to achieve the delivery of air from the equipment to the room can have a great impact on comfort in the room. The capacity of a duct to carry air is affected by the resistance within the duct. Typical materials used to make up ductwork are galvanized steel that can be formed into round or rectangular shapes, foil-faced fiberglass duct board that can be formed into angular shapes in the field, or an insulated flexible fabric round duct. Due to the varying friction factors of the different materials, the type of material must be considered in the design process. Increased friction or duct length adds to the total pressure loss. Galvanized steel material ductwork has a lower friction rate than that of flexible duct due to the smoother inside surface of the steel duct; therefore, the fittings chosen will have a greater impact on friction losses in the system. Because air travels easiest in a straight line, a turn in the ductwork will increase the effective length of the duct by more than just the physical length of the turn. For example, a 90° bend will add as much as 35 additional feet of effective length to a duct run, depending on the construction of the ducts. Effective length values for common fittings are listed in Appendix 3 of ACCA Manual D (Rutkowski 2009).

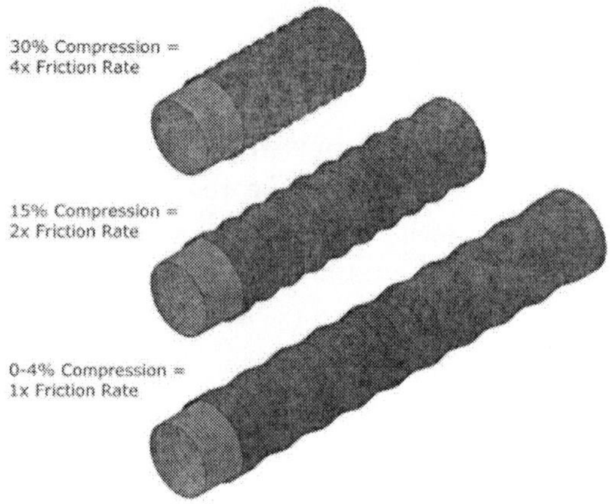

Figure 3. Increased friction rate in compressed flexible duct.

The success of a design in either flexible duct or rigid duct depends upon the faithful execution of the design during installation. The ACCA Standard 5 HVAC Installation Specification (ACCA 2010) was introduced in 2010 as an American National Standards Institute (ANSI) standard to improve the core competencies of contractors and to ensure that quality installations occur. Figure 3 shows how installing flexible duct in its compressed form can increase the friction rate exponentially due to the increased roughness inside the compressed duct.

2.2. Supply Outlets

2.2.1. Impact on Uniformity

When selecting supply outlets, the designer must consider the "occupied zone" of the space to be 2 feet off walls and 6 feet above the floor to avoid drafts. Drafts caused by airflow within the occupied zone can be considered a comfort issue. The amount of airflow that may be perceived as a draft can be relatively low. Air at a velocity of 15 feet per minute may allow smoke to hang in the air, whereas a piece of paper may blow off a desk at an air velocity of 65 feet per minute. The target terminal velocity or the velocity at the useful end of an air stream is between 50 and 75 feet per minute; therefore, if this air stream is directed into the occupied zone, a complaint of a draft is likely.

Supply outlet – The device that introduces supply air to the space.

Pressure loss associated with the supply outlets must be considered when selecting the type of air outlet and must be added to the entire duct run pressure loss.

The proper amount of air is determined by the room-by-room heating and cooling loads. Too little air delivered to a room results in underheating or undercooling, whereas too much air delivered to a room results in overheating or overcooling. To maintain a uniform temperature and to avoid stratification, the supply air must also adequately mix the air in the room.

Mixing of air within the room is a function of the supply outlet and the shape of the air stream entering the room. Figure 4 shows the general concept of a high sidewall supply air delivery pattern, and Figure 5 shows the general concept of a ceiling supply outlet air delivery pattern.

Figure 4. High sidewall supply outlet example.

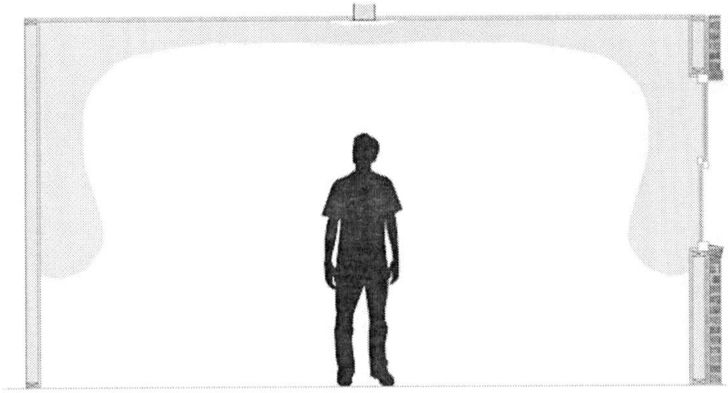

Figure 5. Ceiling supply outlet example.

2.2.2. Throw

By delivering the supply air to the ceiling, the designer is taking advantage of the Coanda effect of air sticking to a surface to increase the throw of the supply outlet. The Coanda effect is a low pressure area that develops between the air stream and the ceiling, allowing the air stream to travel along the ceiling and further increasing the throw.

> **Coanda effect** – The tendency of moving air to attach itself to a surface and flow along it. As air from the supply outlet moves across the ceiling surface, a certain amount of friction occurs between the supply air and the ceiling surface. This resistance to the flow of the supply air pulls the supply air toward the ceiling surface, causing it to stick to the surface.

Throw is the maximum distance the air can reach effectively at a given face velocity and terminal velocity. Air outlets are categorized by the throw as well as the spread (i.e., the width of the air stream), as shown in Figure 6.

The throw of an outlet is more important than the spread because the air must reach into the room as far as possible for good mixing. A narrow spread can be overcome with the use of additional outlets if necessary. It is important that any air exiting multiple outlets does not collide in the room because this will cause the undesired sensation of a draft. If the supply outlets are to be placed perpendicular to each other in a room, then adjustable blades are desirable to be able to direct the air streams so that they do not collide.

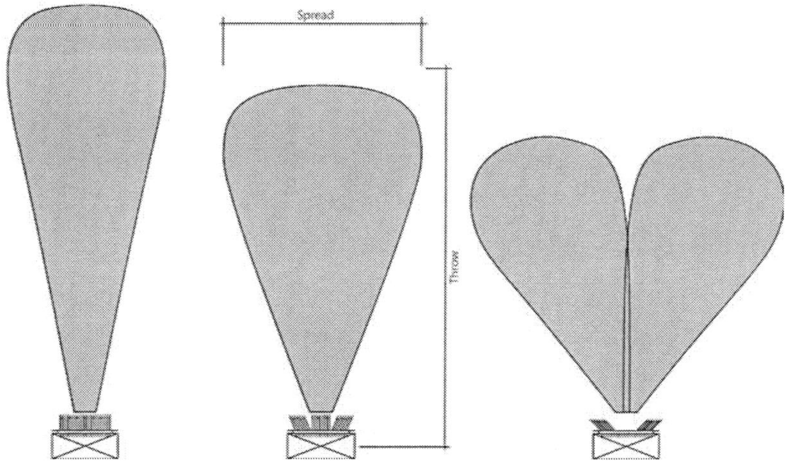

Figure 6. Example throw and spread of air outlets.

Face velocity – The velocity of supply air as it leaves the supply outlet.

Terminal velocity – The velocity of supply air at the end of its effective throw.

Throw – The effective distance that air leaving the supply outlet can reach.

Spread – The width of the air stream leaving the supply outlet as determined by the characteristics of the supply outlet.

2.3. Noise

Noise from the equipment, ducts, or air outlets is an important part of the comfort equation. The placement of the equipment within the conditioned space will increase energy efficiency; however, the placement of the equipment must consider the impact of the equipment noise during operation. Placing equipment close to important noise control areas such as bedrooms must involve consideration of the implications from the equipment noise and vibrations.

The placement of the return air inlet can also have an impact on system noise. A return duct that has a direct connection to the blower motor, as shown in Figure 7, will transfer that blower noise to the occupied space. One way to overcome this issue is by adding radius elbows in the return duct to help isolate the blower noise from the space, as shown in Figure 8.

As air speed increases in duct systems, so does the noise level. Ducts are sized to maintain the maximum velocity of air without adding noise to the room. The type of air outlet chosen and its placement in the room will have an impact on the noise levels in the house. Air that leaves an air outlet at a higher velocity than the outlet is intended to handle will create an undesirable whistling or hissing noise. Improperly placed or selected air outlets can also create a draft in the occupied zone, which is a perceived comfort issue. When selecting an air outlet, the manufacturer's performance data will list a noise criteria (NC) rating based on a very specific set of testing data. The NC ratings are based on laboratory testing of the air outlets. It is important to realize that the listed criteria are for only that specific test situation; however, the data provide a comparative rating. The NC ratings for residential settings are NC 30 or lower. Table 1 shows how the NC ratings apply to familiar settings.

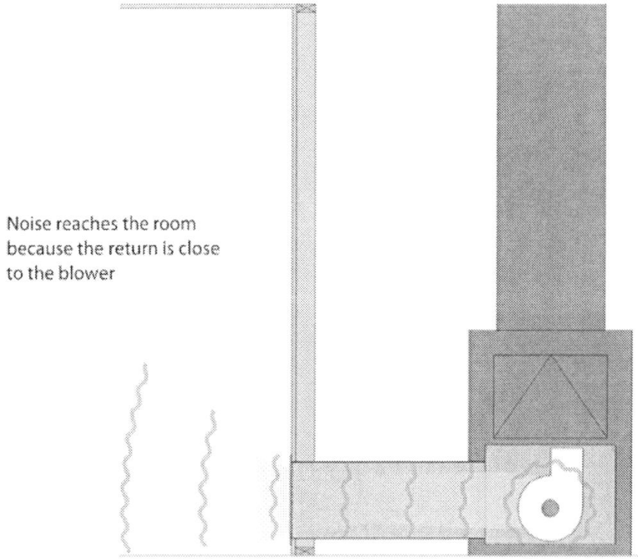

Figure 7. Return placement introduces noise.

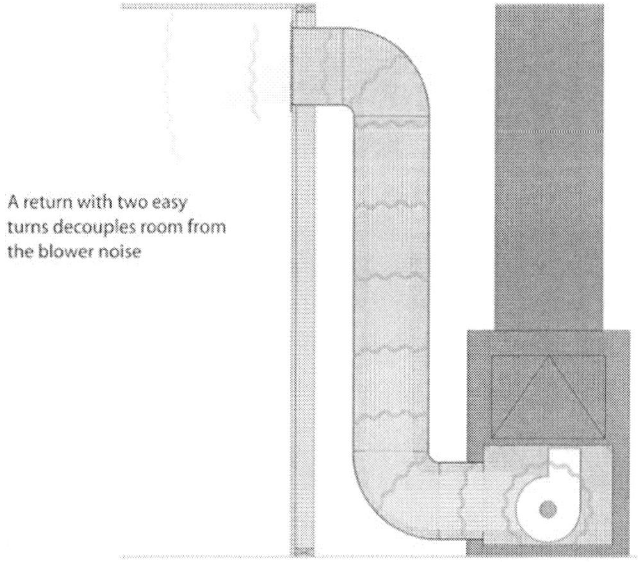

Figure 8. Strategy to reduce return noise.

Table 1. Comparative NC Values

	Communication Environment	Typical Occupancy
<NC 25	Extremely quiet environment; suppressed speech is quite audible; suitable for acute pickup of all sounds.	Broadcasting studios, concert halls,music rooms.
NC 30	Very quiet office; suitable for large conferences; telephone use satisfactory.	Residences, theaters, libraries, executive offices, director's rooms.
NC 35	Quiet office; satisfactory for conference at a 15-foot table; normal voice 10 to 30 feet; telephone use satisfactory.	Private offices, schools, hotel guestrooms, courtrooms, churches, hospital rooms.
NC 40	Satisfactory for conferences at a 6- to 8-foot table; telephone use satisfactory.	General office, labs, dining rooms.
NC 45	Satisfactory for conferences at a 4- to 5-foot table; normal voice 3 to 6 feet; raised voice 6 to 12 feet; telephone use occasionally difficult.	Retail stores, cafeterias, lobby areas, large drafting and engineering offices, reception areas.
>NC 50	Unsatisfactory for conference of more than two or three persons; normal voice 1 to 2 feet; raised voice 3 to 6 feet; telephone use slightly difficult.	Computer rooms, stenographic pools, print machine rooms, process areas.

3. INTEGRATED DESIGN

The HVAC system must be considered in the early schematic design phase to achieve optimal performance in an energy efficient house. During the schematic phase, the design team needs to allocate adequate space for the equipment and ducts while identifying principal potential conflicts between the building's structure and the HVAC system. Decisions made during the early design phase will be critical to the successful performance of the HVAC system.

Locating the HVAC equipment centrally within the house is an early design consideration with many benefits for the performance of the system and implications for the space planning in the house. Locating the equipment centrally will allow for shorter duct runs with similar lengths, which can lead to a better balanced system and improved performance. Centrally located equipment with shorter duct runs also facilitates running ducts to interior walls with the high performance strategy of more efficient high sidewall diffusers aimed at the exterior walls.

An energy efficient house utilizes strategies to keep all ductwork inside the thermal boundaries of the house. Keeping all ductwork inside thermal boundaries will eliminate losses to the outside of the building enclosure but may require the use of soffits that reduce ceiling heights or chases that must be designed with the floor plan flow.

A preliminary layout of the duct system can be made on the floor plan, taking into consideration the performance criteria. By considering a preliminary duct layout, early accommodations can be made in the framing plan as needed.

Floor systems are a commonly used element to run duct systems within the thermal boundary, particularly in multistory houses. Creating chases deliberately when designing the floor plan layout will allow the HVAC system to perform as designed. Considering the location of horizontal and vertical chases early in the design can decrease the level of complexity in the duct runs by stacking the vertical chases and avoiding the reduced efficiencies and increased costs of a complex system layout. These designated chases should be clearly identified and set aside on the construction prints as "reserved" for HVAC. HVAC routing is much less flexible than plumbing and electrical, and the HVAC space should be clearly set aside in design documents. Aligning wall studs with the floor structural members facilitates horizontal to vertical duct transitions.

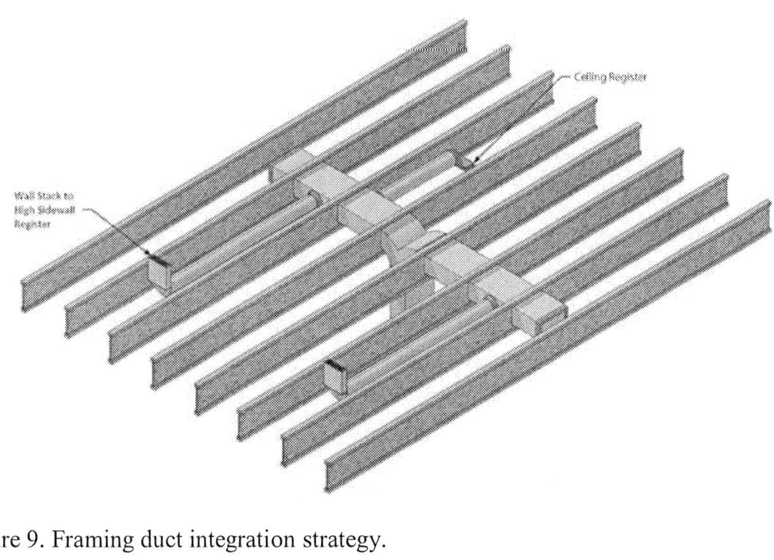

Figure 9. Framing duct integration strategy.

Early in the design stage, beneficial choices in framing materials can be made. As shown in Figure 9, floor joists can be cut to allow the HVAC contractor to route the ductwork both parallel and perpendicular to the floor framing. Following the joist manufacturer's guidance on hole size and spacing is critical to the structural performance of the joist; however, with advance planning and design, these locations can be predetermined, thereby reducing the need for field fitting.

The location of the return air duct system should also be determined early in the design process. Utilizing a central return strategy is a simple and effective way to move air back to the air handler. In locating a central return, it should be placed in a central hallway where it is adjacent to the main living space of the house. Strategies for locating the return air pathways from rooms not open to the central space are discussed in the Return Air Design section of this document.

Some preliminary information about the system airflows will need to be estimated because the size of the return duct system is designed to limit the pressure losses within the ductwork and the size of the return will dictate where it can be placed. The placement of the central return duct is dictated by its size and the location of equipment.

The location of the return also affects the location of the filter, and the filter location will have a direct impact on the usability of the system. Placing filters in accessible locations will allow homeowners to easily replace the filters. In a house where the equipment is located in the basement and is easily accessible, placing the filter at the equipment return air inlet may be an appropriate choice. However, when equipment is located in the attic and is less accessible, a filter located at the return grille may be a better choice.

The use of building cavities as air passageways to deliver air to or from the HVAC system, such as wall cavities between studs and wallboard or floor cavities where material is used to box in floor joists, is not appropriate in an energy efficient house. A fully ducted return system will be easier to seal and will have better airflow characteristics than building cavities. Return air pathways that leak will draw air from unintended places in the house and can lead to undesirable pressure differences.

4. Design Principles

The distribution system must be designed to get the most out of the available air from the blower power that was selected based on the equipment manufacturer's performance data to meet the estimated heating and cooling loads. The blower capacity of the equipment selected will deliver a specific amount of air against a specific amount of system pressure. The resistance of the airflow within the distribution system must match the static pressure against which the equipment is able to perform. The ducts, fittings, supply outlets, and other accessories installed in the air stream all add to the pressure loss between the equipment and the outlets. The return system also adds to this system pressure loss and must be included in the design.

4.1. Pressure Drop

To ensure proper performance of the system, the pressure loss of each component of the system must be accounted for and subtracted from the available static pressure of the equipment. A worksheet such as the External Static Pressure Worksheet shown in Figure 10 can help the designer keep track of the various components with their associated pressure losses in the system. The sum of all the pressure losses will equal the total pressure drop, and the pressure drop must match the available static pressure of the equipment.

Static Pressure – A resistance to airflow due to the airflow restrictions created by coils, filters, elbows, grilles, duct friction, etc.

Available Pressure – The difference between the static pressure and the pressure against which the selected equipment can deliver the cubic feet per minute of air. This value is what is available in the design for duct and supply outlet pressure losses.

4.2. Room-by Room Supply Requirements

Using the heating and cooling load room-by-room estimates, the volume of conditioned air needed to satisfy the load of each room must be calculated. Software packages will perform this task for the designer, but it is important to understand what parameters influence the values. ACCA Manual D (Rutkowski 2009) describes a two-step method for determining the room cubic feet per minute based on the equipment's design blower capacity and the heating or cooling load. First a heating factor (HF) or cooling factor (CF) is determined by dividing the design blower cubic feet per minute of the equipment by the total heating or cooling load. The HF or CF value is then multiplied by the room load to determine the heating or cooling cubic feet per minute required to meet the room load.

HF (or CF) = Design Blower Cubic Feet per Minute ÷ Total Heating (or Cooling) Load

ACCA Manual D procedures select the highest airflow for each room and rely on seasonal adjustments of the balancing dampers to achieve the seasonal airflows. In an energy efficient house where ducts are in conditioned space, access to balancing dampers may not be possible. In addition to the potential for balancing dampers to not be accessible in concealed ductwork, this strategy relies on the occupant making the adjustments at least twice a year. Where balancing dampers cannot be avoided by careful duct sizing to airflows, the placement of the damper near the boot for fine tuning will help avoid accessibility issues.

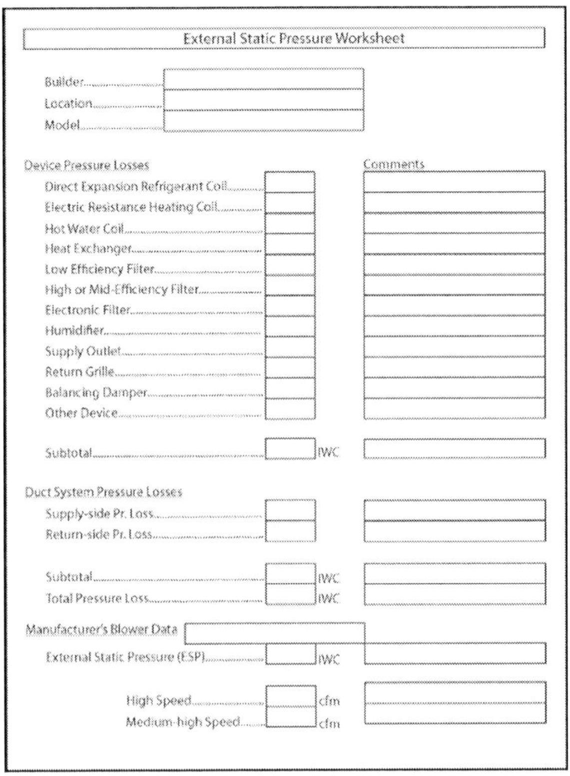
External Static Pressure Worksheet

Builder
Location
Model

Device Pressure Losses
Direct Expansion Refrigerant Coil
Electric Resistance Heating Coil
Hot Water Coil
Heat Exchanger
Low Efficiency Filter
High or Mid-Efficiency Filter
Electronic Filter
Humidifier
Supply Outlet
Return Grille
Balancing Damper
Other Device

Comments

Subtotal IWC

Duct System Pressure Losses
Supply-side Pr. Loss
Return-side Pr. Loss

Subtotal IWC
Total Pressure Loss IWC

Manufacturer's Blower Data
External Static Pressure (ESP) IWC

High Speed cfm
Medium-high Speed cfm

Figure 10. Pressure worksheet.

Table 2 shows the room-by-room loads with the corresponding cubic-feet-per-minute requirements. The design cubic-feet-per-minute values used are the average values between the heating and cooling requirements. The Orlando House floor plan in Figure 11 shows how rooms such as the family room and the kitchen and nook can be combined based on their proximity and the performance characteristics of the ceiling supply outlet chosen.

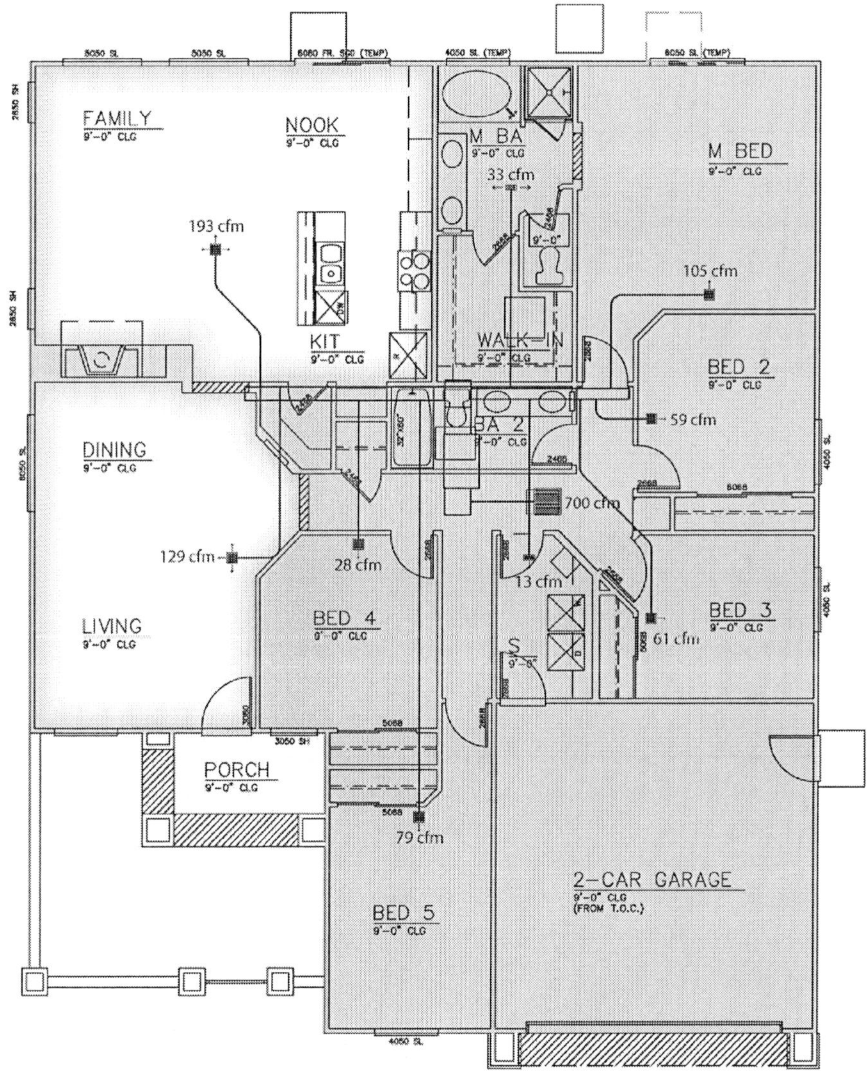

Figure 11. Orlando House floor plan.

Table 2. Orlando House Room-by-Room Loads

	Peak Heating Load			Peak Cooling Load		Actual Supply Airflows
Room	**Btu/h**	**cfm**	**Btu/h**	**cfm**	**Heating**	**Cooling**
Family	4,459	132	4,055	137	193	193
Kitchen/Nook	2,094	62	1,628	55	0	0
Master Closet	0	0	0	0	0	0
Master Bath	1,250	37	858	29	33	33
Bath 2	0	0	0	0	0	0
Master Bedroom	3,614	107	3,049	103	105	105
Bedroom 2	1,655	49	2,013	68	59	59
Bedroom 3	1,723	51	2,101	71	61	61
Laundry	405	12	444	15	13	13
Bedroom 4	1,013	30	770	26	28	28
Living/Dining	4,256	126	3,907	132	129	129
Hallway	0	0	0	0	0	0
Bedroom 5	3,175	94	1,865	63	79	79
TOTALS	23,644	700	20,689	700	700	700

4.3. Return Air Design

A well-designed return air strategy is critical for the performance of the HVAC system in an energy efficient house with lower airflow requirements to meet the load (see Figure 12). The return air must "see" a clear path from every room that has a supply outlet, with the exception of bathrooms or kitchens due to the potential for spreading odors through the house.

A direct return from each room to the equipment is not desirable when trying to minimize duct lengths and pressure losses in the return system. Rooms without an adequate return air path will become pressurized, which will impede supply airflow and could result in comfort complaints. Typical door undercuts (1/2 to 3/4 inch) do not allow adequate return volume and are not appropriate for an energy efficient house. Door undercuts are specifically not approved in ACCA Manual D (Rutkowski 2009).

A low resistance in the return air path is defined as having less than 0.05 inch water column (iwc) pressure loss. The pressure loss for a filter at the return grille must also be considered in the return air stream. A target value for return capacity is two times the volume of the total supply air with an airflow velocity within the return at less than 500 feet per minute and the net free area of the grille sized 1.5 times the cross-sectional area of the return duct.

Utilizing a central return strategy is a simple and effective way to get air back to the air handler. In locating a central return, it should be placed in a central hallway where it is adjacent to the main living space of the house. Figure 13 shows the location of the central return in the ceiling of the Orlando House plan. Each room that has a door to close it off from the central space with the exception of the laundry and the hall bath utilizes an over-the-door transfer as a clear path from that room back to the central return.

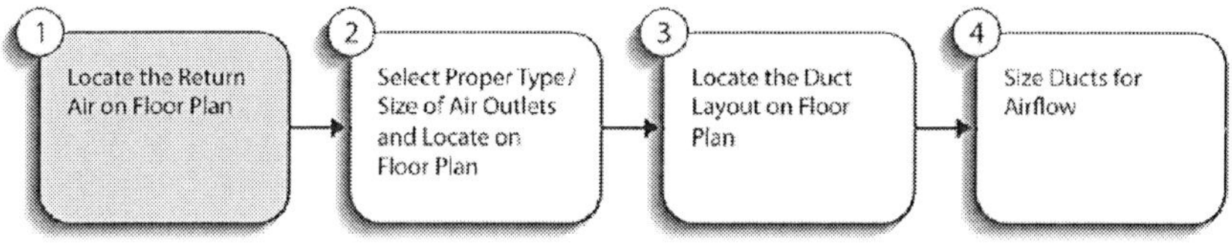

Figure 12. Locate the return first.

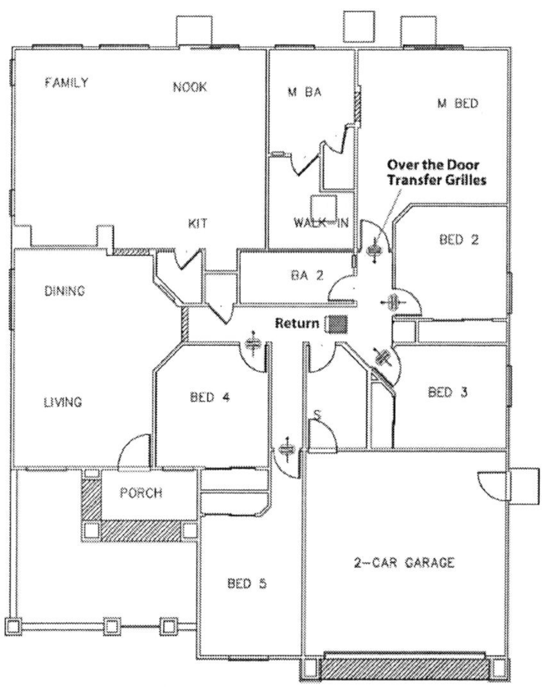

Figure 13. Orlando House return location.

Figure 14 of the Chicago House plan shows a similar central return strategy except with a high sidewall return grille in a chase leading to the basement where the equipment is located.

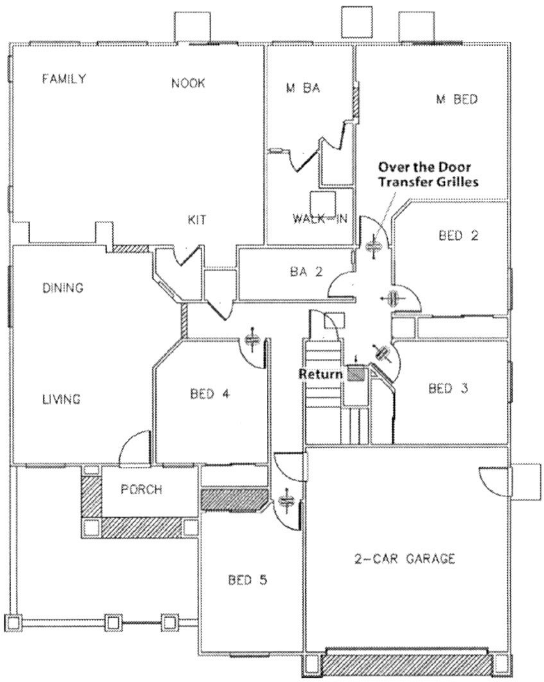

Figure 14 Chicago House return location.

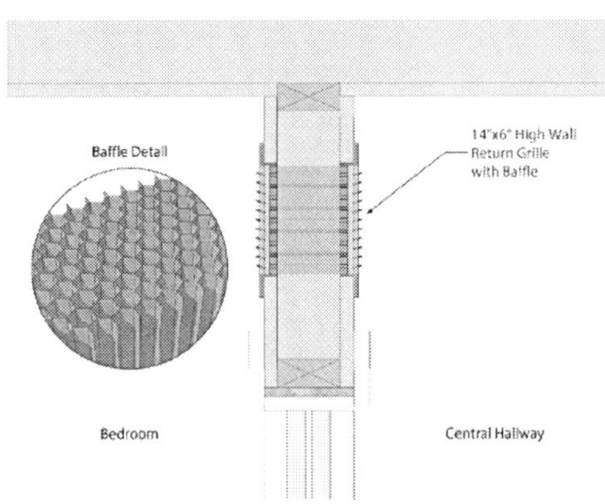

Figure 15. Over-the-door transfer.

An over-the-door transfer as shown in Figure 15, a high/low through-the-wall transfer as shown in Figure 16, or an attic transfer duct as shown in Figure 17 with adequate net free area are all appropriate solutions to provide a return air path without compromising privacy in the room.

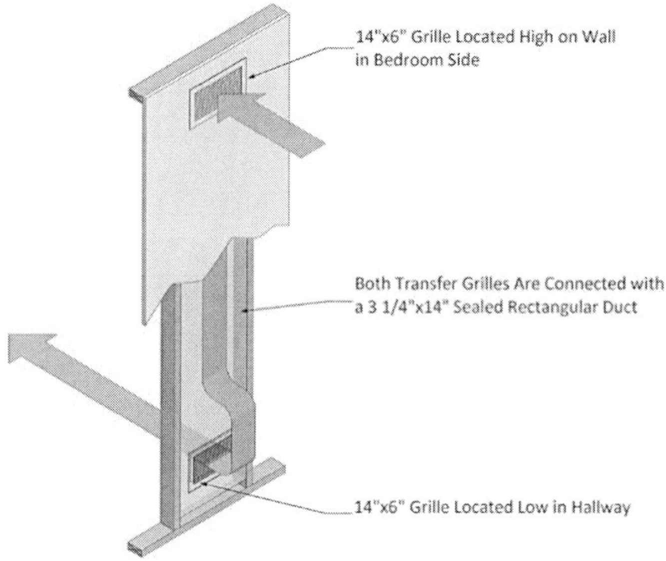

Figure 16. High/low through-the-wall transfer grille.

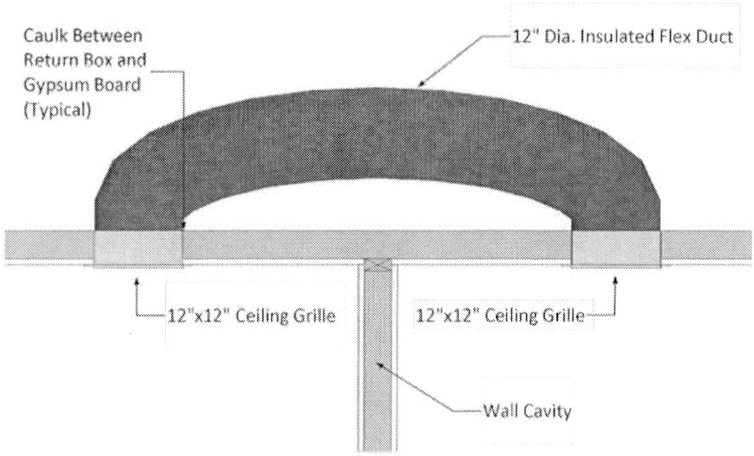

Figure 17. Attic transfer duct.

Even with the proper net free area and sizing of the return duct, airflow can be restricted by turbulent air at the entrance of the return duct, causing poor performance and noise issues. In the example shown in Figure 18, the bottom of the return grille is flush with the top of the duct. This scenario will create turbulence at the back of the duct entrance that limits flow into the top of the grille, reducing the total return flow. By extending the return riser up one-third the area of the return grille as shown in Figure 19, the airflow within the riser is smoothed out, allowing for better flow back to the equipment.

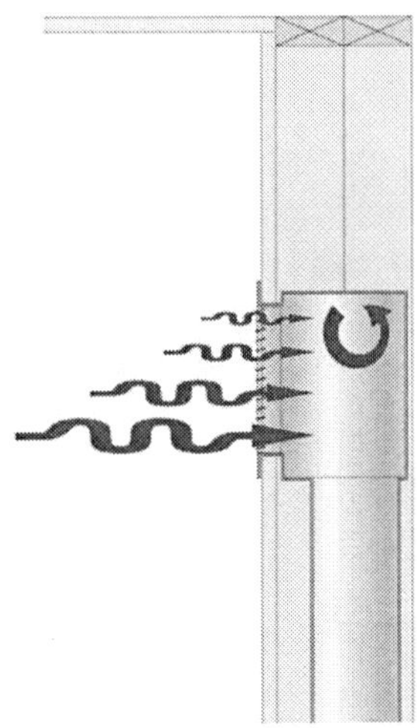

Figure 18. Poor return grille placement.

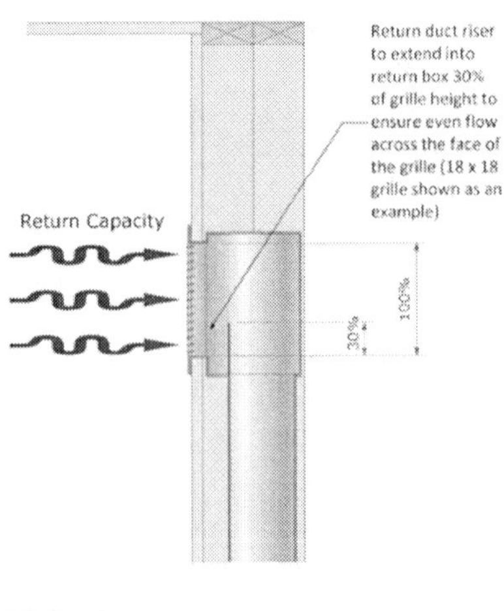

Figure 19. Good design, high return.

4.4. Supply Air Outlets

The selection and placement of the supply air outlets are critical to the comfort in the space (see Figure 20). The air must be delivered in a manner that mixes the supply air with the room air without introducing unacceptable noise or causing the sensation of a draft on the occupant. The manner in which the air is distributed in the room is the function of the shape and size of the air outlet. Placement of the air outlet is as important to the comfort in the room as is the type of air outlet that is selected.

Historically, due to poor performing windows and lack of wall insulation, supply outlets were located at the perimeter of the house to "wash" the thermal enclosure with conditioned air. In an energy efficient house, rooms can be adequately conditioned from high sidewall or ceiling registers located toward the interior of the house, with the airflow directed toward the exterior walls. Floor registers at the perimeter should not be used because they are susceptible to being covered or blocked by furniture, making them completely ineffective. A room with only a single supply outlet that has been blocked by furnishings can be uncomfortable to an occupant. Where a floor register is unavoidable, it must be sized to the lower airflows of an energy efficient house.

A distribution strategy for the lower airflows required to meet the lower loads of an energy efficient house includes locating supply ducts on the interior walls and using more efficient outlets that throw the air to the exterior walls. This allows for better mixing with the room air. Ducts and supply outlets should be sized for the following purposes:

- Maintain the air velocity as high as possible without creating noise problems.
- Promote mixing of the air in the room.
- Limit stratification.

The strategy utilized for the Chicago House is a high sidewall throwing outlet, while the Orlando House uses a curved blade ceiling outlet. Both strategies direct the air outward across the room and high above the occupants' heads, where the air can mix with the room air to deliver the desired comfort.

There are four general steps to selecting the proper air outlet, as shown in Figure 21. Although it may appear that these steps are sequential, steps 1, 2, and 3 are iterative and may need to be performed several times before the correct supply outlet is selected. The designer must continually strike a balance between desired throw characteristics and pressure drop during the air distribution design process.

The required cubic feet per minute for the room being serviced by the supply outlet is a critical input to the selection process as well as the face velocity, throw, spread, pressure loss, mounting location, and NC.

The target face velocity for the high performance strategy is between 700 and 900 feet per minute. For the high performance strategy with a high interior sidewall outlet, the ideal scenario would be a throw that just reaches the opposing exterior wall at a terminal velocity of 75 feet per minute or has a throw that is 2 feet greater than the distance to the wall at a terminal velocity of 50 feet per minute.

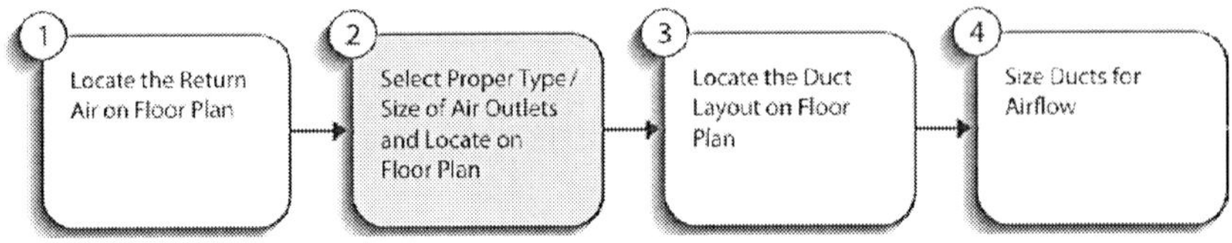

Figure 20. Select and size air outlets.

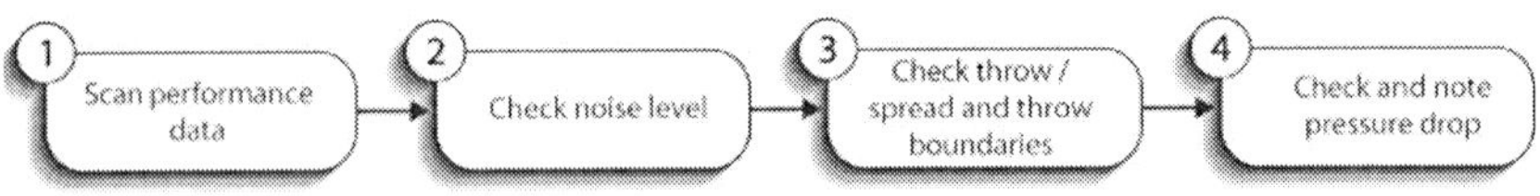

Figure 21. Air supply outlet selection.

Face Velocity		400	500	600	700	800	900	1000	1100
Pressure Loss		.010	.016	.022	.031	.040	.050	.062	.075
8 x 4	cfm	65	80	100	110	130	145	160	175
Ak .160	Throw	6.5	8.0	10.0	11.0	13.0	15.0	16.0	18.0
10 x 4	cfm	80	100	120	140	160	180	200	220
Ak .202	Throw	7.0	9.0	11.0	13.0	14.0	16.0	18.0	20.0
12 x 4	cfm	100	120	145	170	195	220	245	270
Ak .244	Throw	8.0	10.0	12.0	14.0	16.0	18.0	20.0	22.0
14 x 4	cfm	115	145	170	200	230	255	285	315
Ak .286	Throw	8.5	11.0	13.0	15.0	17.0	19.0	22.0	24.0
12 x 5	cfm	125	155	190	220	250	280	310	345
Ak .312	Throw	9.0	11.0	14.0	16.0	18.0	20.0	22.0	25.0
10 x 6	cfm	125	155	190	220	250	285	315	345
Ak .314	Throw	9.0	11.0	14.0	16.0	18.0	21.0	23.0	25.0

Ak= net area in square feet

Figure 22. Example supply outlet performance data.

The performance data for a register are published by the manufacturer in a data table such as the one shown in Figure 22.

The performance data are often published at a terminal velocity value of 75 to 100 feet per minute. If the listed throw at the 75-feet-per-minute terminal velocity is the throw necessary for the room, then these data are acceptable; however, it is likely that it will be necessary to interpolate the throw distance at 50 feet per minute. The procedures for doing these calculations are best handled using the calculations for non-isothermal jets as described in the ASHRAE 2009 Handbook—Fundamentals (ASHRAE 2009b)."

Other considerations that are not directly performance related but should be considered are availability, price, material, and appearance. Availability of low-cost high sidewall or ceiling supply outlets to satisfactorily meet the low air volumes of an energy efficient house with good throw characteristics can present a challenge in design. More options are becoming available in manufacturers' residential lines, but it may be necessary to search light commercial options to find appropriate outlets with the desired performance characteristics.

The additional costs of a higher performing curved blade ceiling outlet or high sidewall outlet can often be offset by a reduction in the number of outlets and the associated ductwork. The Orlando House family room and kitchen and nook area as shown in Figure 23 would require three or four typical residential supply outlets run to the exterior walls in a traditional design, whereas in this example, one three-way curved blade outlet in the center of the room supplies the necessary airflow.

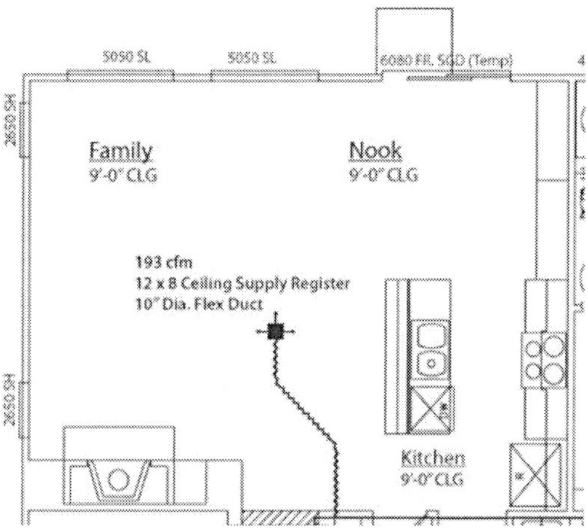

Figure 23. Orlando House family room and kitchen and nook area.

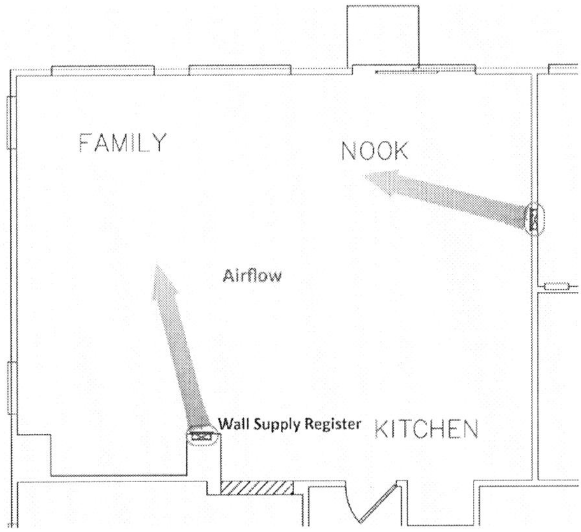

Figure 24. Chicago House family room and kitchen and nook area.

The same space in the Chicago House is serviced by a pair of one-way throw high sidewall supply outlets on interior walls aimed toward the exterior wall (see Figure 24). Selecting an outlet with adjustable blades will allow for the air to be directed into the room, avoiding a collision mid-room and the potential for the sensation of a draft.

4.5. Duct Layout

From the early integrated design work and the room-by-room airflows, a preliminary layout of the ductwork can be completed on the floor plans (see Figure 25). Figure 26 shows the layout and cubic-feet-per-minute requirements as well as the outlet locations and types for the Orlando House. A combination of one-way, two-way, and three-way ceiling supply outlets is used, along with insulated flexible duct.

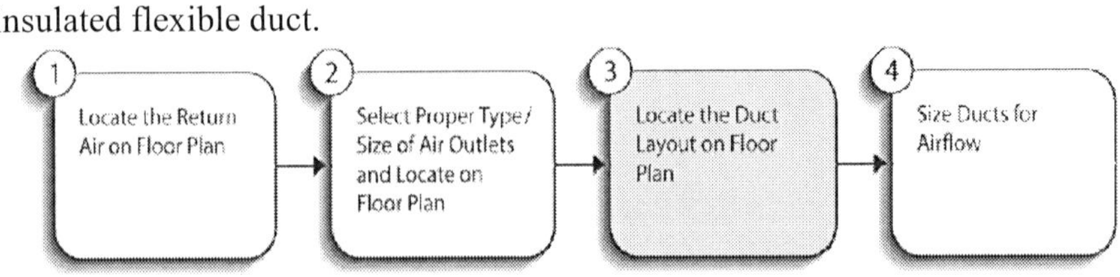

Figure 25. Duct layout.

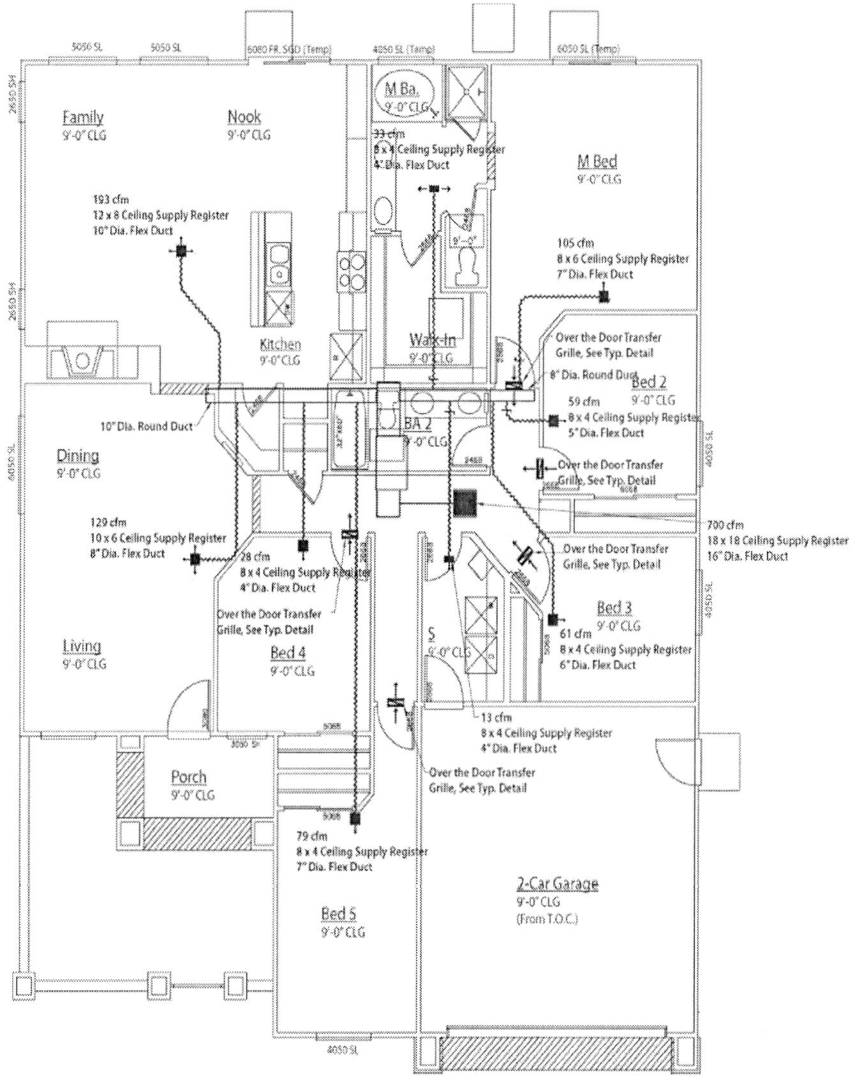

Figure 26. Orlando House layout.

The Chicago House as shown in Figure 27 uses oval wall stacks and all galvanized steel ducts to bring the air up from the basement to the high sidewall supply outlet locations. Care is taken to avoid bringing ductwork out of the conditioned space and into the attic. The basement with the equipment and duct location is shown in Figure 28. Two supplies and one return service the unfinished basement area.

4.6. Duct Size

The duct size is selected relative to the amount of air that must pass through the ductwork (see Figure 29). The variables that define the performance of the duct airflow are the cubic feet per minute of air into the room based on the heating and cooling loads, the friction loss in the duct, the cross-sectional area of the duct, and the average velocity of the air moving in the duct.

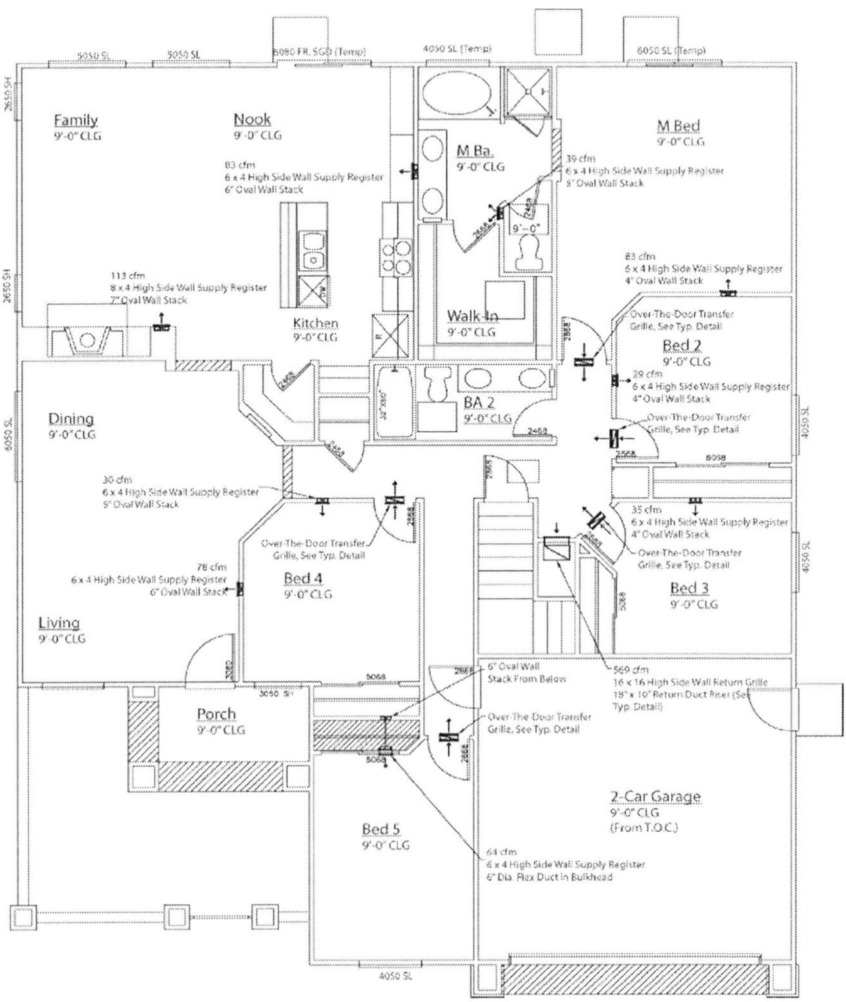

Figure 27. Chicago House main floor.

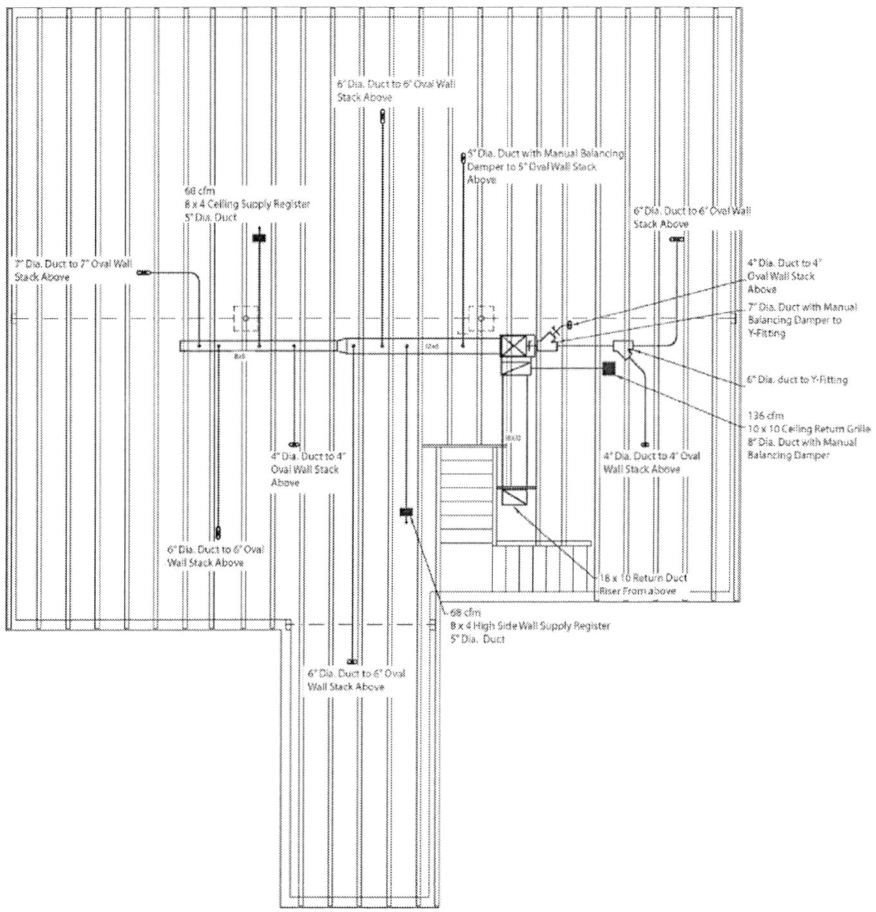

Figure 28. Chicago House basement layout.

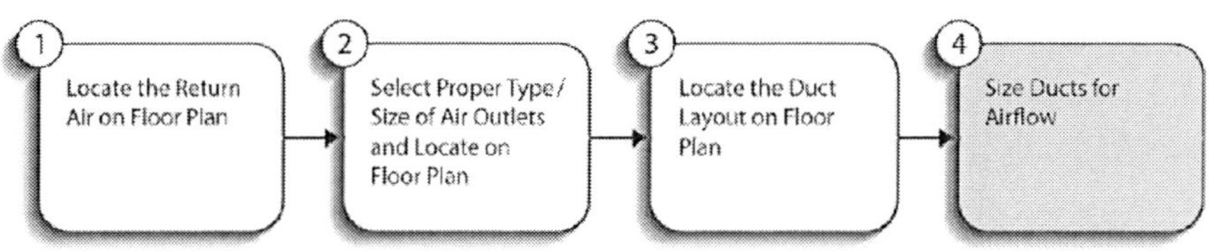

Figure 29. Size ducts for airflow.

ACCA Manual D procedures calculate the friction rate based on the worst case or "critical path" duct run that has the longest effective length (Rutkowski 2009). The critical path friction rate is then used for the design of all other duct runs, even though the actual friction rate may be lower. This method slightly oversizes ducts with shorter runs, and the ACCA Manual D standard of care requires the use of balancing dampers in every duct run to adjust the airflow to each room.

Friction rate – The pressure loss between two points in a duct system that are separated by a specific length. The units for friction rate are inches water column per 100 feet (iwc/100 ft).

Effective length – The length of duct used for design purposes that includes the straight length of duct added to all the fitting equivalent lengths.

Equivalent length – The amount of duct length added to a duct run by fittings. Often the equivalent length of a fitting can be equal to or longer than the straight duct lengths due to the additional friction of turning air.

ACCA Manual D acknowledges that in an energy efficient house where ducts are in conditioned space, access to balancing dampers may not be possible; therefore, dampers cannot be installed. The ACCA Manual D standard of care neither endorses nor condemns the practice of designing duct sizes based on the individual friction rate and flow and only offers caution regarding the complexity of this design method and the potential for inadequate performance if noise criteria are ignored.

The *ASHRAE 2009 Handbook—Fundamentals* (ASHRAE 2009a) describes the procedures to evaluate the friction rate at each duct run and then to size the duct accordingly to achieve a "self-balancing" system. Calculating the friction rate of each duct in an effort to accurately select the correct duct for a balanced system without dampers does require more rigorous calculations because most commercial software is based on the ACCA duct design methodology. In most cases, averaging the airflows between the heating and cooling requirements for self-balancing systems is a reasonable compromise. However, where there is a larger disparity in the loads, the designer should evaluate the room-by-room loads and may want to favor a heating or cooling airflow for rooms with unique conditions, such as sunrooms, rooms over garages, and rooms with significantly longer runs than most others.

The calculations for a self-balancing system will be more complex. The size of duct calculated may be non-standard, requiring careful design consideration in choosing the next larger or smaller standard size. Also, close attention must be given to avoid excessive velocity in smaller ducts. Duct performance is defined by the following four independent variables:

- Duct diameter
- Flow rate (cubic feet per minute)
- Velocity
- Friction rate.

To manage all of the variables and calculations of the ASHRAE method, the use of a spreadsheet for calculations is critical. Table 3 shows how the pressure losses in each supply run are tracked based on a duct velocity of 427 feet per minute, standard duct diameters, the required cubic feet per minute per room, the roughness factor of galvanized steel, and the effective length of the duct with fittings included.

Table 3. Chicago House Supply Pressure Loss

Duct Run	Room	Take-off Dia. (in.)	Duct Length Dia. (in.)	Branch (cfm)	Duct Length Material	Effective Length	Pressure Loss
1	Family	7	7	113	Galvanized Steel	99	0.070
2	Living/Dining	6	6	78	Galvanized Steel	92	0.067
3	Basement Rear	5	5	68	Galvanized Steel	69	0.073
4	Bedroom 4	4	4	30	Galvanized Steel	68	0.054
5	Bedroom 5	6	6	64	Galvanized Steel	85	0.052
6	Kitchen/Nook	6	6	83	Galvanized Steel	80	0.068
7	Basement Front	5	5	68	Galvanized Steel	60	0.070
8	Master Bath	5	5	39	Galvanized Steel	66	0.037
9	Master Bedroom	6	6	83	Galvanized Steel	58	0.052
10	Bedroom 3	4	4	35	Galvanized Steel	49	0.049
11	Bedroom 2	4	4	29	Galvanized Steel	37	0.025
					Minimum	37	0.025
					Average	69	0.056
					Maximum	99	0.073

Table 4. Master Bedroom Duct Run

Element	Equivalent Length	Adjusted Manual D Fitting EL	Manual D Fitting Number	Manual D Reverence Velocity	Diameter (in.)	Flow Rate (cfm)	Velocity (fpm)	Friction Rate (in./100 ft)	Pressure Loss(iwc)
Branch Take-off	50	20.0	2-P	900	8	149	427	0.045	0.00899773
Branch Length	31.25	31.3			8	149	427	0.073	0.02295476
Elbow	5	6.2	11	400	8	149	427	0.073	0.0045551
Diffuser Boot	60	23.9	4-Z	900	8	149	427	0.045	0.01079727
Subtotals		81.34779							0.047305

Default equivalent lengths are calculated based on the velocity of 900 feet per minute. As an example, the effective length for the master bedroom is calculated for the design velocity of 427 feet per minute with the data shown in Table 4.

$$EL_x = EL^* (V_x / V)^2$$

where

EL = effective length

V = velocity

CONCLUSION

The higher performing house has a lower load, and consequently less air volume is needed to condition the space. This presents both challenges and opportunities when designing the air distribution system. Challenges arise when using air outlets sized by “rule of thumb” that do not have the throw needed to provide air mixing in the room to achieve the desired comfort results. At the same time, the energy efficient house design process offers opportunities to integrate the HVAC system into the design earlier in the process.

Early consideration of the HVAC system in the whole house design can lead to an efficient and comfortable house. Trying to force a system late in the design process may end up with undesirable consequences.

The equipment selected to meet the heating and cooling loads has a specific capacity of air delivery against a specific static pressure. Each component in the air distribution system has an associated pressure loss that must not exceed the available static pressure of the equipment.

Lower volumes of air must be delivered to the space in a manner that will provide good mixing with the space air to provide comfort. Careful consideration of the air outlet strategy and a full duct design are critical to the HVAC system delivering the comfort in an energy efficient house, whether it is new construction or an energy upgrade retrofit. Both system noise and noise at the air outlet are important comfort considerations in the air distribution system design.

The return air system to bring air back to the air handling equipment is critical to the comfort levels within the house. Because the return air ducts are typically much larger than the supply ducts, the placement of the return air system is to be considered early in the design process.

In the entire HVAC design process, heating and cooling loads are dependent on the construction and location of the house, while the equipment selection is dependent on the loads, and the air distribution system is dependent on both the loads and the equipment capacity. The success of a design in either flexible duct or rigid duct depends on the faithful execution of the design during installation. The ACCA/ANSI Standard 5 (ACCA 2010) has been established to ensure that quality installations occur.

References

[1] ACCA. ACCA Standard 5 HVAC Quality Installation Specification. Arlington, VA: Air Conditioning Contractors of America. 2010.

[2] Air Diffusion Council. ADC Flexible Duct Performance and Installation Standards, 5th Edition. Schaumburg, IL: Air Diffusion Council. 2010.

[3] ASHRAE. *ASHRAE 2009 Handbook—Fundamentals*, Chapter 21, Duct Design. Atlanta, GA: American Society of Heating, Refrigerating and Air-Conditioning Engineers. 2009a.

[4] ASHRAE. *ASHRAE 2009 Handbook—Fundamentals*, Chapter 20, Space Air Diffusion. Atlanta, GA: American Society of Heating, Refrigerating and Air-Conditioning Engineers. 2009b.

[5] Burdick, A. (2011a). *Strategy Guideline: Accurate Heating and Cooling Load Calculations*. Washington, DC: U.S. Department of Energy. www.nrel.gov/docs/fy11osti/51603.pdf.

[6] Burdick, A. (2011b). *Strategy Guideline: HVAC Equipment Sizing*. Washington, DC: U.S. Department of Energy.

[7] Rutkowski, H. *Manual D—Residential Duct Systems*, 3rd Edition. Arlington, VA: Air Conditioning Contractors of America. 2009.

[8] Rutkowski, H. *Manual J—Residential Load Calculation*, 8th Edition. Arlington, VA: Air Conditioning Contractors of America. 2006.

[9] Rutkowski, H. *Manual S—Residential Equipment Selection*. Arlington, VA: Air Conditioning Contractors of America. 1995.

[10] Rutkowski, H. *Manual T—Air Distribution Basics for Residential and Small Commercial Buildings*. Arlington, VA: Air Conditioning Contractors of America. 1992.

In: Guidelines for Improved Duct Design … ISBN 978-1-62081-443-7
Editors: L. M. Collins, D. E. Martinez

Chapter 2

Measure Guideline: Sealing and Insulating Ducts in Existing Homes*

R. Aldrich and S. Puttagunta

Definitions

AC	Air conditioning
ACCA	Air Conditioning Contractors of America
AHU	Air handler unit
BPI	Building Performance Institute
CAZ	Combustion appliance zone
ccSPF	Closed-cell spray polyurethane foam
CFM	Cubic feet per minute
EPA	Environmental Protection Agency
HVAC	Heating, Ventilating, and Air Conditioning
IAQ	Indoor Air Quality
IECC	International Energy Conservation Code
IRC	International Residential Code
MSDS	Material Safety Data Sheet

* This is an edited, reformatted and augmented version of Office of Energy Efficiency and Renewable Energy, U.S. Department of Energy, dated December 2011.

SMACNA	Sheet Metal and Air Conditioning Contractors National Association
wrt	with reference to

FOREWORD

Heating and cooling losses from forced-air ducts can result in high energy costs, lead to thermal comfort problems, and–in some extreme situations–result in serious health and safety concerns. Reducing air leakage and conductive losses from ducts can be a straight-forward way to reduce energy use and improve comfort in homes.

The authors hope that this document is useful to a wide audience: builders, remodelers, HVAC contractors, home performance contractors, homeowners, etc. Some of the procedures presented here, however, require specialized equipment or expertise. In addition, some alterations to duct systems may require a specialized license. Persons implementing duct system improvements should not go beyond their expertise or qualifications.

This document begins with a discussion on potential cost and performance benefits of duct sealing and insulating (p.3). It continues with a review of typical duct materials and components (p. 6). The overall procedures for assessing and improving the duct system are:

1. Inspect the home and HVAC systems (p. 13): Identify possible hazards and assess duct system materials and conditions.
2. Combustion safety testing (p. 14): Unless all appliances are sealed-combustion, it is important to check combustion equipment for proper operation before and after sealing, because sealing ducts can alter pressure balances in a home.
3. Air system diagnostics (p. 18): Check for adequate system flow and, if appropriate, measure duct leakage levels.
4. Seal the ducts (p. 22)
5. Insulate the ducts (p. 37)

Note that this is not intended as a guide for designing, installing, or performing major retrofits to HVAC systems. It is certainly true that a great many duct systems are poorly designed and installed, and reconfiguration may lead to improved efficiency and comfort.

ACKNOWLEDGEMENTS

The authors would like to acknowledge the funding and support of the U.S. Department of Energy's (DOE) Building America Program. This guide is the product of a collaborative effort. Special thanks to David Lee of the U.S. Department of Energy and Ren Anderson and Cheryn Engebrecht of the National Renewable Energy Laboratory (NREL).

PROGRESSION SUMMARY

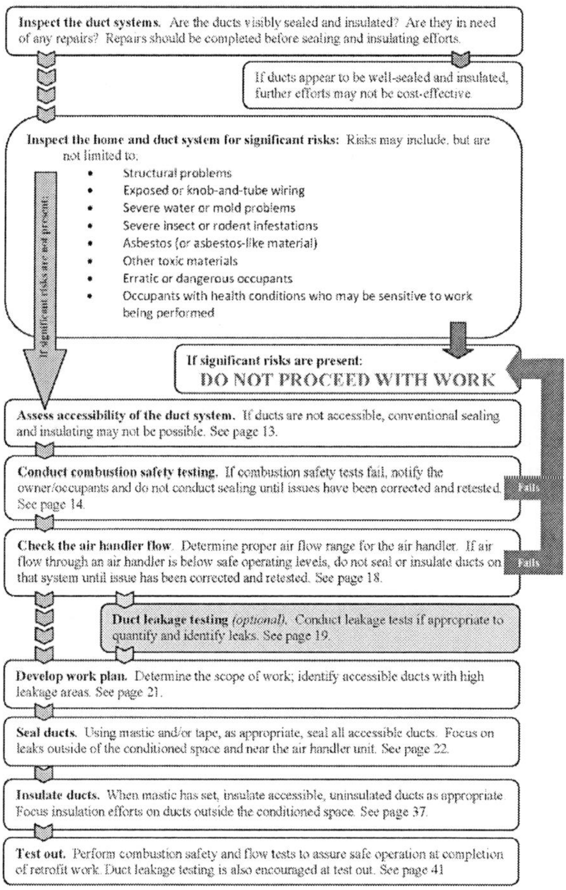

1. Introduction

Ducts carry conditioned air between a central air handler and rooms throughout a home. In an ideal system, all of the supply air leaving the air handler would be transferred by the duct system to various points of use in the home and ultimately be discharged through supply registers within the rooms. Similarly, all of the return air in an ideal system would be drawn from the appropriate return air grilles, through return ducts, and back to the furnace or air handler. Historically, this has not been the case. According to the EPA ENERGY STAR® program, the typical home has 20% duct leakage. In older homes, it is not uncommon for duct leakage to account for 30%-50% of the total system air flow. Much higher leakage rates are possible in very poorly sealed systems. Ideally, duct leakage would be negligible, but a goal of less than 5% of the total system air flow is reasonable for newly installed systems. For existing duct systems, when parts of the duct system are not accessible, going from 30% leakage to 5% leakage is often not realistic. Ducts should be sealed as tightly as is practical, and this level varies tremendously from system to system.

Major Duct Retrofits

If major HVAC retrofits are planned as part of a project, the following resources may be helpful:

- Air Conditioning Contractors of America (ACCA) *Manual J (8th edition), Manual D, Manual S, Manual T, Manual RS, and Standard 5: HVAC Quality Installation Specification*
- EPA document "ENERGY STAR Qualified Homes, Version 3 (Rev. 02) HVAC System Quality Installation Contractor Checklist"
- ASHRAE *Handbook of Fundamentals 2009*
- Sheet Metal and Air Conditioning Contractors National Association (SMACNA) *Residential Comfort System Installation Standards Manual*
- Building Industry Institute's *Procedures for HVAC System Installation Scope of Work*
- ASHRAE Standard 152 - Method of Test for Determining the Design and Seasonal Efficiencies of Residential Thermal Distribution Systems

When ducts are located in (or just passing through) unconditioned spaces such as attics or crawlspaces, a significant amount of energy–and heating and cooling capacity–can be lost. If ducts are located within conditioned parts of the home, duct leakage might not be a major energy liability, but still may lead to comfort issues.

Even when ducts are in conditioned space, duct leakage can prevent the proper amount of air from reaching the intended spaces. This can lead to comfort problems because portions of the home are over- or under-conditioned.

Duct leakage can also lead to pressure imbalances within homes. Such imbalances not only can affect comfort and efficiency, but can also impact health and durability. In homes with some types of combustion equipment, for example, large return duct leaks in a basement system can cause negative pressures which, in turn, can interfere with proper draft. Under these conditions, exhaust gases can be sucked into the home. Other risks of pressure imbalances include build up of moisture (and associated problems like mold) in parts of buildings.

Wherever ducts are located, duct leakage should be minimized for optimal efficiency, comfort, and durability in the home. While not addressed in building codes until recently, the 2009 International Energy Conservation Code (IECC) mandates duct leakage to outdoors be no more than 8 CFM_{25} per 100 ft_2 of conditioned floor area (see section 403.2 of the 2009 IECC for details). High performance building programs, such as EPA's ENERGY STAR Qualified Homes and USGBC's LEED® for Homes, require total duct leakage to outdoors not be more than 4 CFM_{25} per 100 ft^2 of conditioned floor area.

Similarly, duct insulation levels strongly impact distribution system performance. The most obvious benefit of insulating ducts is reducing the heat loss from (during heating) or heat gain to (during cooling) the ducted air. This is again more critical when ducts are located in unconditioned spaces, but there may be reasons to insulate ducts in conditioned spaces as well. In humid climates, proper duct insulation can prevent condensation on the cool surface of the ducts (and possible related moisture problems). In long duct runs, adding insulation will reduce energy loss as the air travels from the air handling unit to the supply registers. This will result in more desirable discharge air temperatures from registers, maximizing system capacity and potentially improving occupant comfort.

Note that this guide is only intended as a resource for sealing and insulating existing duct systems in homes; this is not intended as a guide for designing, installing, or performing major retrofits to HVAC systems. It is certainly true that a great many duct systems are poorly designed and installed, and reconfiguration may lead to improved efficiency and comfort.

2. Cost and Performance

While duct sealing can be a simple and effective way to improve home HVAC performance, the costs and benefits vary tremendously from home to home and across various climates. This section presents a framework and examples for assessing costs and benefits of duct sealing and insulation.

2.1. Energy Savings

NREL's Residential Efficiency Database (NREL Duct Sealing 2011) presents duct sealing options to reduce leakage from 30% to 15% or 6% of total system airflow. To demonstrate the energy savings of this air sealing, CARB referred to modeling used to develop the "*Retrofitting Home Analysis Dashboard*" (CARB). This demonstration shows the effects of various retrofit measures on a typical 1970s home in various climates.

In both climates assessed here (Houston and Boston), the home is 1,647 ft2 with three bedrooms. In Houston, the home is on a slab with ducts in the attic (total duct area of 445 ft2 and insulated to R-6). In Boston, the home was modeled with uninsulated ducts located in an unconditioned basement (same duct area as the Houston model). The modeled homes have poor insulation–typical of the 1970s–and old HVAC equipment with effective SEER of 8 and AFUE of 76%. For more details on the homes, see CARB's *Retrofitting America: A 1970s Home Energy Efficiency Analysis*. Table 1 shows the energy and cost impacts of duct sealing in Houston.

Table 1. Modeling results on the reference home in Houston with various levels of duct leakage

Duct Leakage to Outdoors	Annual Energy Consumption		Annual Heating and Cooling Cost	Annual Savings
	Cooling [kWh]	Heating [therms]		
30%	11,547	470	$1,922	$0
15%	10,258	434	$1,719	$203
6%	9,056	404	$1,541	$381

Table 2. Modeling results on the reference home in Boston with various levels of duct leakage

Duct Leakage to Outdoors	Annual Energy Consumption		Annual Heating and Cooling Cost	Annual Savings
	Cooling [kWh]	Heating [therms]		
30%	3,195	1160	$1,893	$0
15%	3,048	1088	$1,767	$126
6%	2,931	1035	$1,677	$216

In a Houston climate, because of the dominant cooling load and ducts located in the hot attic, duct sealing can save $200-$400 per year in HVAC costs. In the Boston example, largely because many ducts are located in the basement, savings are more modest.

Energy costs in these examples were taken from 2005 RECS (EIA 2005). In Houston, electricity costs were $0.1165/kWh and gas costs $1.101/therm. In Boston, energy costs were $0.13/kWh and $1.189/therm.

Of course, the savings achieved by duct sealing depends on many, many factors. These examples simply demonstrate that duct leakage can indeed account for a significant portion of heating and cooling costs.

Energy savings from duct insulation varies with the initial state of duct insulation, the duct locations, and how much of the duct system one can effectively insulate. In the examples above, going from entirely uninsulated ducts to R-6 ducts will save approximately $100 per year in both climates (Boston and Houston). Going to R-8 duct insulation provides modest additional savings of $5 per year.

2.2. Cost

The examples above show potential for $100 - $400 in annual savings from reducing duct leakage. The next likely question is: how much does it cost to seal ducts? As with savings, the costs to seal ducts vary tremendously. Duct location, type, duct area, and accessibility to ducts vary from home to home. As described above, labor costs typically dominate duct sealing efforts; the sealants themselves are usually a much less significant portion of the sealing costs. Contractor rates, therefore, influence total sealing costs tremendously.

Taking these caveats into account, NREL has compiled cost ranges for duct sealing (NREL Duct Sealing). These costs are summarized in Table 3 along with cost ranges for the example home described above.

Table 3. Duct sealing costs from NREL' s ' ' National Residential Efficiency Measures Database' ' . These costs are applicable to both example homes discussed above

Leakage Reductions	Cost per sq.ft. of Duct Area	Total Cost for Example Home
30% to 15%	$0.64 - $3.00	$285 - $1,335
15% to 6%	$0.64 - $3.00	$285 - $1,335
30% to 6%	$1.30 - $6.00	$579 - $2,670

NREL has also compiled typical price ranges for duct insulation (NREL Duct Insulation); these ranges are shown in Table 4.

2.3. Cost-Performance Trade-Offs and other Solutions

As discussed above, the energy impact of duct losses (by leakage and/or conduction) from ducts that are located within the building envelope are usually minimal. In fact, many modeling tools consider ducts located within conditioned space to have no net energy losses–regardless of leakage or insulation conditions. While tight, well-insulated ducts within the envelope are better for overall performance and comfort reasons, energy impacts are minimal. Moving ducts within conditioned space (or expanding conditioned space to encapsulate ducts) is sometimes a viable retrofit strategy.

Aside from moving ducts into conditioned space, and short of complete removal and/or replacement of a forced-air HVAC system, there is no good substitute for duct sealing and insulating. As home heating and cooling loads are reduced (from envelope improvements, high-efficiency equipment, etc.), the relative impact of duct losses will change, but these losses can still be a liability with respect to energy, equipment capacity, and comfort.

Table 4. Duct insulation costs from NREL' s ' ' National Residential Efficiency Measures Database' ' . These costs are applicable to both example homes discussed above

Duct Insulation	Cost per sq.ft. of Duct Area		Total Cost for Example Home	
R-6	$0.70	- $1.50	$311	- $668
R-8	$0.82	- $1.80	$365	- $801

2.4. Non-Energy Benefits

There are three potential benefits from reducing duct losses: comfort, HVAC capacity, and IAQ/combustion safety

2.4.1. Comfort

The comfort benefits are very straight-forward. If heat (or "coolness") is lost from a duct before the air reaches the register, the room where that register is located will receive less heating or cooling. In large homes, exterior rooms with long, leaky duct runs can be much less comfortable than inner rooms with shorter ducts runs (and possibly where thermostats are located).

2.4.2. HVAC Capacity

If a home requires four tons of cooling, and duct losses amount to one additional ton (easily possible with duct leakage over 20%), a five-ton AC system would be needed to meet the loads under design conditions. If a four-ton system was installed, there would likely be comfort problems on design days.

These capacity issues are often compounded by extreme conditions, especially when ducts are located in the attic. On hot summer days, attic temperatures can easily reach 130°F or 140°F. Supply ducts that carry 50°F - 55°F air experience at least an 80°F temperature differential; return leaks can draw in 130°F attic air–this obviously requires much more power to cool than indoor air at around 75°F.

Minimizing duct losses can allow for smaller equipment–especially with respect to cooling. If equipment capacity is already stretched, reducing or eliminating duct losses can allow more of the heating and cooling capacity to be delivered to the home.

2.4.3. IAQ and Combustion Safety

Duct leakage can cause pressure imbalances in homes. Excessive supply leaks can cause a home to be depressurized; this, in turn, can cause drafting problems and can potentially draw combustion products into the home. Depressurization in hot, humid climates can also lead to moisture problems if humid outdoor air is drawn into cooler envelope assemblies.

To minimize combustion safety risks, combustion safety testing should be conducted both before and after duct sealing is performed.

3. Ductwork Components

3.1. Duct Materials

This section discusses the three most common duct materials: sheet metal, fiberglass duct board, and flexible duct (typically referred to as "flex duct"). Each material has advantages and disadvantages and can require different approaches for insulating and air sealing. A duct distribution system may utilize a combination of these materials, such as sheet metal trunks off the air handling unit and flex duct branches to supply the individual registers.

3.1.1. Sheet Metal

Galvanized steel metal is a very common duct material and is required in some jurisdictions. Round and rectangular shapes are the most common, but ovals and other shapes are sometimes used. There is also a wide array of sheet metal fittings that connect sections of duct (elbows, transitions, takeoffs, reducers, etc.).

Screws are typically used to connect straight round ducts to collars, elbows, T's, Y's, reductions, boots, and other fittings. SMACNA recommends using at least three #8 sheet metal screws spaced equidistant. If seams, gaps, and connections are not sealed before insulation, substantial leakage can result.

Figure 1. Sheet metal supply plenum sealed with mastic. A flexible, fireproof canvas collar was installed to provide sound/vibration attenuation.

Figure 2. Fiberglass duct board ducts in an attic.

3.1.2. Fiberglass Duct Board

Duct board is a rigid, dense, fiberglass sheet; the outside typically has a fiber-reinforced foil facing which acts as a vapor barrier. Duct board is usually sold in large sheets (e.g. 4' x 10') with thicknesses of 0.75" - 2". The fiberglass provides thermal insulation (approximately R-4 per inch) as well as sound attenuation. Duct board is typically cut, folded, and shaped into rectangular sections of duct. Duct board is primarily used for plenums, trunks, and junction boxes; smaller branch ducts can be sheet metal or, more commonly, insulated flex duct. As with sheet metal, the most common areas for leakage are at collars, transitions, or connections to other materials.

3.1.3. Flex Duct

Insulated, flexible duct consists of an interior polymer tube (4-24" in diameter) supported by a spiral, metal coil. This inner liner is surrounded by fiberglass insulation (1"-3" thick), and then an outer polymer liner, which usually incorporates a low-emissivity foil facing that provides a radiant barrier as well as a vapor barrier. Flex duct is relatively inexpensive, and it comes pre-insulated. Because it is flexible, flex duct can be run longer distances—with several bends and turns—without the need for additional fittings or connections (flex duct typically is sold in 25-foot lengths). With fewer transitions and fittings, flex duct can be very air-tight. As with other duct materials, leakage problems still occur at connections at either end of the duct and may not be apparent because gaps are covered by insulation. Connection and sealing of flex should always be done at the inner liner.

HELPFUL TIP

Flex duct needs to be properly supported to minimize pressure losses in the distribution system. SMACNA recommends that support spacing not exceed 5 feet and a maximum sag of less than ½ inch per foot of spacing between the supports.

Compared to other types of duct material, flex duct is less durable. It can be twisted, compressed, torn, or otherwise disturbed, resulting in leaks. Flex ducts also have higher resistance to air flow (especially when the duct is not fully extended), so slightly larger ducts are needed to carry the same volume of air.

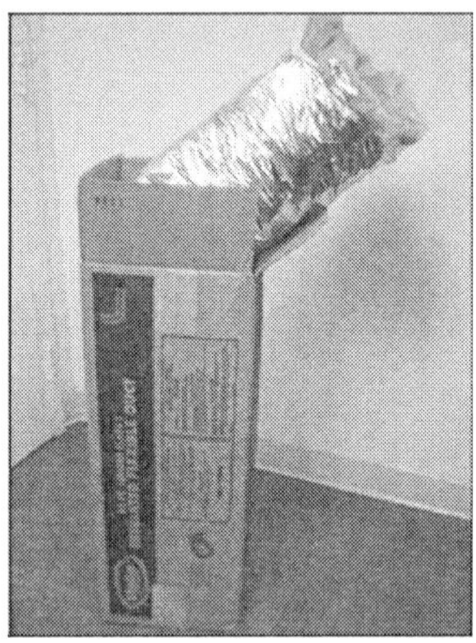

Figure 3. Box containing 25 feet of compressed insulated flex duct.

Figure 4. Insulated flex duct in the attic of an existing home.

3.1.4. Building Cavities

In many homes, both old and new, HVAC systems make use of building cavities to move air, especially return air. While this can result in some installation cost and space savings, it is not recommended. One reason is that building cavities are very prone to air leakage: gaps in framing and/or drywall are very common, and many building cavities have electrical and plumbing penetrations. Using building cavities as ductwork on the return air side can also result in pollutants entering the air stream and being distributed throughout the home.

One of the most common types of building cavities used as an air duct is the floor joist as a "panned" return. A panned return consists of a floor joist bay where "panning" has been installed across the bottom of two or more joists. In cases where the return grille is located in a wall, an uninsulated wall cavity may be used to connect to a panned return.

Figure 5. Floor cavity being used as return pathway. Note the electrical penetrations into this "panned" return.

Figure 6. A "panned" return where the building cavity is used to carry air.

3.2. Sealants

Ironically, most conventional "duct tape" is not an acceptable product for duct sealing. Proper duct sealants are classified according to two Underwriters Laboratories standards:

- UL-181A, "Standard for Closure Systems for Use With Rigid Air Ducts"
- UL-181B, "Standard for Closure Systems for Use With Flexible Air Ducts and Air"

These standards generally describe the air sealing of new duct systems; bringing older systems up to these levels may not be possible or practical. At a minimum, however, duct sealing products should meet the requirements of the appropriate standard above; these products will be labeled "UL181A" or "UL181B". Many materials are approved for both A and B, i.e. for sealing both rigid and flexible ducts. Language from Section M1604.4.1 of the 2009 International Residential Code (IRC) may be appropriate to incorporate into duct sealing scopes of work (see sidebar).

2009 IRC, M1601.4.1 Joints and Seams

"Joints of duct systems shall be made substantially airtight by means of tapes, mastics, liquid sealants, gasketing or other approved closure systems. Closure systems used with rigid fibrous glass ducts shall comply with UL181A and shall be marked 181A-P for pressure-sensitive tape, 181 A-M for mastic or 181 A-H for heat-sensitive tape. Closure systems used with flexible air ducts and flexible air connectors shall comply with UL 181B and shall be marked 181 B-FX for pressure-sensitive tape or 181 B-M for mastic. Duct connections to flanges of air distribution system equipment or sheet metal fittings shall be mechanically fastened. Mechanical fasteners for use with flexible nonmetallic air ducts shall comply with UL 181B and shall be marked 181 B-C…."

ACCA Standard 5: HVAC Quality Installation Specification and the companion Checklist are useful resources for HVAC contractor scopes of work if a major overhaul of the existing duct system is being performed. Another excellent resource is a document published by the U.S. Department of

Energy's Building Technologies Program called "Air Distribution System Installation and Sealing: Proper Duct Installation Increases Efficiency."

3.2.1. Tapes

While conventional duct tape is typically not appropriate, there are many tapes with UL-181A and UL-181B ratings. Most of these have aluminum foil backing instead of the cloth backing of conventional duct tape. Many of these rated tapes, however, are still prone to failure over time (Sherman et al. 2000). Thin foil tapes are also not able to conform to curves and irregular shapes of ducts; these tapes can also easily be punctured and torn (e.g. by sharp edges of ducts). Also, for proper adhesion, duct surfaces must be clean and dry before tapes are applied.

One type of tape does offer more durability—butyl duct tapes. While expensive, these tapes have a thicker, more rugged adhesive that resists tearing and conforms more easily to irregular surfaces. Whatever tape is used, follow the manufacturer's instructions for surface preparation, applied pressure, and application temperature.

Figure 7. Two UL rated duct tapes (Butyl tape, left roll, and Aluminum foil tape, right roll).

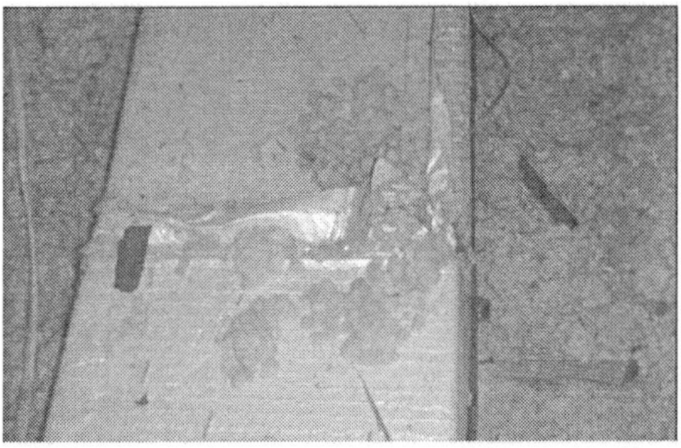

Figure 8. Aluminum tape falling off of duct board ducts.

Figure 9. A gallon of mastic and mesh tape.

3.2.2. Mastic

Most duct mastic is a water-based paste that dries to form a semi-rigid, waterproof coating around duct joints and seams. Mastic is typically sold in 1, 2, or 5-gallon pails as well as caulk tubes. When cured, mastic will maintain some elasticity, and it can tolerate temperature changes and vibrations associated with the air distribution system. Mastic is often used in conjunction with fiber mesh tape , such as products used for drywall seams. This mesh can provide reinforcement, especially for larger gaps (greater than 1/8") or penetrations.

Many home energy experts consider mastic superior to tapes in most situations. Mastic will conform to irregular, sharp or rough edges where tape may rip or leak. The chief complaint about mastic is usually that it is inconvenient to apply. Although mastic can be messier than tape, the ability to apply it with a tool, caulking gun, or by hand make it the best option for properly sealing hard-to-reach cracks.

Figure 10. A duct connection being sealed using fiberglass mesh tape and mastic being applied with a paint brush.

Armed with inexpensive disposable gloves and/or single use paint brushes, experienced installers can perform most duct sealing jobs neatly, more quickly, and often more effectively with mastic than with tape alone.

Certain mastics can also be applied via spray. Whatever mastic is used, follow the manufacturer's instructions for surface preparation, application method, and application temperature.

3.2.3. Aerosol Sealant

In the 1990s, researchers at Lawrence Berkeley National Laboratory developed a method to seal ducts from the inside with aerosol sealants blown through the duct system. With registers blocked, the sealant would leak only through seams or gaps in the duct system; and the aerosol sealant would build up around the edges of the gaps, gradually closing in the holes and sealing the ducts. While this system cannot seal large holes or disconnects in the duct system, it can be especially effective where ducts cannot be accessed for traditional sealing methods (i.e. mastic or tape). Currently, this technology is offered commercially under the trade name Aeroseal.

Figure 11. Duct board sealed with mastic.

Figure 12. A duct board trunk with flex duct branches sealed with mastic.

3.3. Insulation

Insulation is typically installed over ductwork to reduce conductive and radiant heat transfer through the duct material and to prevent condensation (during cooling). The 2009 IECC requires that all ductwork located in attics be insulated to a minimum of R-8 and all other ductwork located in unconditioned spaces to be insulated to a minimum of R-6. Older homes may have R-4 or no insulation on ductwork. All replacement duct insulation materials should meet the requirements of 2009 IRC Section M1601.3.

3.3.1. Fiberglass

Fiberglass is the most common duct insulation material. As discussed above, duct board and insulated flex duct are manufactured with fiberglass insulation to provide some level of thermal resistance (R-value is typically labeled on the duct surfaces). Uninsulated flex ducts are also available, although these are not recommended for use as supply ducts nor in unconditioned spaces (return ducts within conditioned space may be an acceptable application for uninsulated flex). Uninsulated flex ducts are sometimes found in systems that were installed or retrofitted by a layperson. In cases where uninsulated flex duct is found, insulation strategies similar to those discussed below for use on sheet metal ductwork can be utilized.

When handling fiberglass insulation, wear protective clothing (gloves, dust mask, safety glasses, and coveralls) to avoid potential skin and eye irritation.

Most sheet metal ducts are uninsulated or wrapped on the outside with fiberglass insulation blankets, sometimes called “duct wrap”. These sheets (typically 2-4 feet wide) are 1-4” thick and include a vapor barrier on the outside, often with a foil radiant barrier. The insulating value (R-value) of fiberglass duct wrap is between 2.5 and 4 ft2hr°F/Btu per inch (depending on the specific product and density). Newer sheet metal ducts that have interior, insulating duct liners are now available, but these liners are not generally appropriate for retrofit applications.

3.3.2. Closed-Cell Polyurethane

Foam Insulating ductwork is currently not a common application for closed-cell spray polyurethane foam (ccSPF), and it is even less common in retrofit applications. While spray foam can provide excellent insulation and air sealing, most codes require that spray foam insulation be covered by an ignition barrier. With ducts located in attics, this can be accomplished by "burying" spray-foamed ducts beneath at least 1.5" of fiberglass insulation. The 2009 IRC addresses this strategy in section M1601.3.

4. Health and Safety

Before describing duct diagnostics, sealing, or insulating procedures, it is important to understand basic health and safety information. Prior to beginning any work, inspect the home and systems to determine if the duct improvement work can be performed safely. It may be appropriate to refer to standards such as the Building Performance Institute's (BPI's) "Technical Standards for the Building Analyst Professional" or other protocols.

The contents of this section are provided as suggestions. Contractors should refer to appropriate local codes, regulations, professional standards, and common sense as the situation warrants.

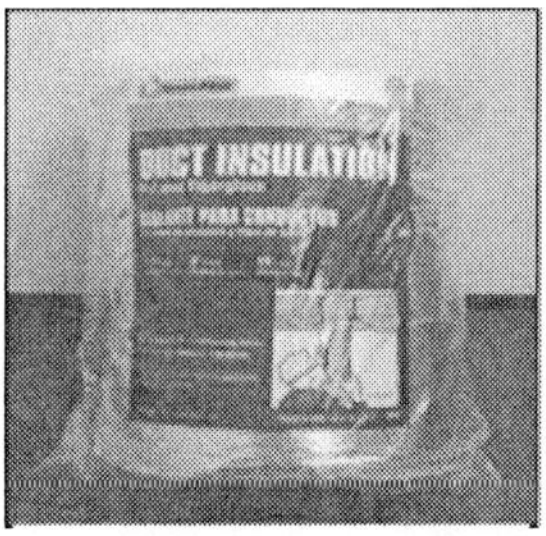

Figure 13. Duct wrap.

Figure 14. Ducts insulated with closed-cell spray foam in an attic. Fiberglass insulation will be blown over the ducts, providing the necessary ignition barrier.

4.1. Inspect the House and HVAC Systems

- Structural Issues: If the building is damaged or structurally unsound, duct sealing work should not be pursued until other repairs are completed.
- Duct Status: Ducts that are disconnected, damaged, or designed poorly can result in poor HVAC performance. If duct systems are to be modified or improved, this should be done before sealing and insulating efforts.
- Asbestos, Mold, and Other Contaminants: If there is an existing mold problem in the building or in the ductwork, or if possible asbestos-containing material is present, the conditions should be documented and assessed to determine if testing and sealing can safely be completed. If testing and sealing cannot be safely completed, the homeowner should be instructed to have the problem(s) addressed by a qualified professional before continuing with duct sealing work.
- Health/Safety of the Occupants: If there are people living in the home with severe medical conditions, the situation should be evaluated to ensure that the duct testing and sealing will not cause adverse health conditions. This could include stirring up dust, blowing cold air into the house, shutting off systems to complete work during extreme weather, etc.
- Health/Safety of the Worker: If ductwork is located in an attic or other overhead space, caution must be taken to prevent damage to the ceilings in the occupied spaces below. Typically, this requires careful navigation along the ceiling structural members, either ceiling joists or truss bottom chords. Insulation often makes it difficult to identify solid footing, which can lead to damage in ceiling sheetrock below. In most homes, roofing nails penetrate the roof sheathing and extend through the underside of the decking. Where head clearance is low, protective head gear should be worn to prevent cuts or scrapes from the nails. Similarly, workers should wear gloves, goggles, masks, and protective clothing to protect themselves from airborne insulation fibers and dust that are released when ductwork is disturbed. Work should only be undertaken by people comfortable working in constrained spaces and familiar with safe practices for access in attics or other overhead spaces.
- Make note and be aware of other hazards: toxic materials, solvents, exposed or knob-and-tube wiring (have an electrician evaluate), non-

IC rated can light fixtures, insect/rodent infestations, etc. Proceed with work only if it can be done safely.

Asbestos

Asbestos can be problematic if the microscopic fibers become airborne. If inhaled, significant health problems can result.

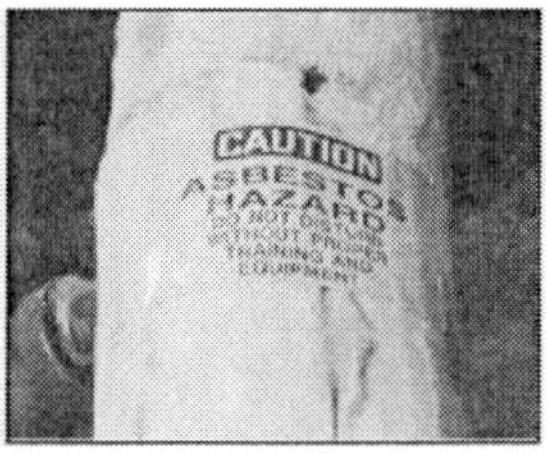

Mold

Mold spores can develop on porous surfaces, in the presence of food (like wood and paper), oxygen, moderate temperatures (typically over 65°F, but if enough water is present, down to 40°F), and sufficient water.

Carbon Monoxide

Carbon Monoxide (CO) is invisible, odorless, and can be deadly. CO poisoning deprives your body's cells of oxygen.

Figure 15. Ceiling damage as a result of a misplaced foot while working on ducts in an attic.

4.2. Combustion Safety Testing

After the initial inspection of the home and systems, it is important to assess combustion appliances in the home and any potential safety issues associated with them. Sealing ducts can result in different pressure patterns within a home, which can potentially alter combustion venting and draft. A brief overview of recommended safety precautions is provided here. It may be appropriate to consult more detailed program guidelines or to refer to BPI's "Technical Standards for the Building Analyst Professional".

4.2.1. Measure Carbon Monoxide (CO) Concentrations

Measuring ambient levels of carbon monoxide in the home is simple and can identify potentially hazardous situations. If ambient CO levels are extremely high, work should not proceed until the problem is corrected. It may also be appropriate to measure CO levels in the flue or exhaust from appliances (usually a different instrument is required). If CO concentrations in the exhaust are high, it may be appropriate to have a qualified contractor adjust or tune the equipment to correct the problem.

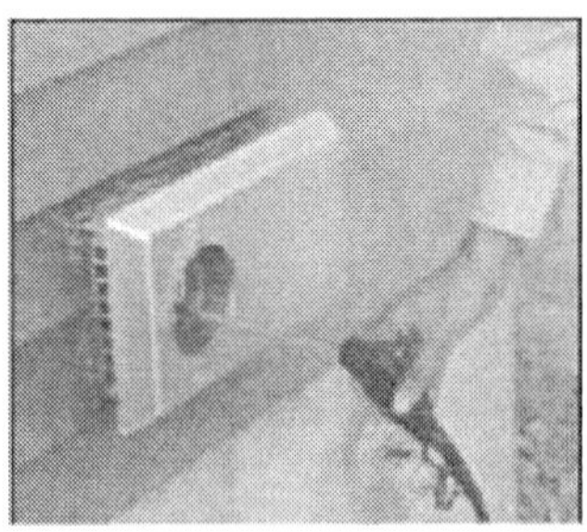

Figure 16. Testing CO levels at the exhaust flue of sealed combustion furnace.

Table 5. Exposure limits for carbon monoxide (CO)[1]

Agency / Organization	Maximum CO Concentrations
Occupational Safety and Health Administration (OSHA)	50 ppm as an 8-hour time-weightedaverage
National Institute for Occupational Safety and Health (NIOSH)	35 ppm as an 8-hour time-weighted average AND 200 ppm at any time
American Conference of Governmental Industrial Hygienists (ACGIH)	25 ppm as an 8-hour time-weightedaverage for a 40-hour work week.

4.2.2. Spillage Tests

"Spillage" is the term used to describe the situation when a proper draft is not established in the flue pipe or chimney; the flue gases come "spilling" back down the flue and into the house. Spillage may be caused by pressure imbalances within the house and/or improper venting. Since flue gases can contain significant amounts of moisture and carbon monoxide, spillage can result in dangerous and unhealthy conditions and should be remedied before attempting any modifications to the home or ductwork.

Check for spillage by turning the appliance on and - using a smoke stick, glass, or mirror—verify that there is negative draft around the draft diverter (i.e. flue gases are flowing up–not down–the flue) that is established within one minute of the burner firing. It is not unusual for some appliances (especially water heaters) to spill temporarily on startup. This is because the chimney needs to heat up before a proper draft is established. Spillage that occurs under normal building operating conditions should be addressed immediately. This is a serious health and safety risk for the occupants. Spillage may also be checked under simulated "worst-case" conditions as described in Section 4.2.4 below.

4.2.3. Measure Draft

Draft is measured by taking a pressure measurement in the flue pipe approximately 12" downstream of the draft diverter. Draft should be measured under steady state conditions (usually after 5-10 minutes of operation). Acceptable draft is dependent on the outside temperature and height of the flue stack. Typical ranges are shown in Table 6.

Figure 17. Spillage testing of an atmospherically vented, natural gas water heater.

Table 6. Typical draft requirements for combustion equipment from BPI' s ' ' Technical Standard for the Building Performance Professional.' '

Outdoor Temperature (T_{out}, °F)	Minimum Acceptable Draft (Pa)
<10	-2.5
10-90	(T_{out} / 40) - 2.75
>90	-0.5

4.2.4. Worst Case Depressurization

It is often recommended that combustion safety tests be conducted under "worst case" conditions. Since testing only gives us a snapshot of how the system is performing on a particular day under the current conditions, it does not necessarily assure us that the system might not fail at another time under different conditions. Using the worst-case approach, contractors are better assured that combustion appliance failures will not occur after they leave the house.

Identifying the CAZ

A combustion appliance zone is any room/space within a home that contains an appliance that burns fuel (wood, natural gas, propane, oil, etc.)

To set up worst case conditions, first identify the combustion appliance zone (CAZ) and set up a manometer to measure the pressure in the CAZ with reference to (wrt) outside. Make changes to the house configuration using the following checklist, and check the CAZ pressure wrt outside at each step. Keep making changes until you have achieved the most negative pressure possible in the CAZ.

Checklist for Establishing Worst-Case Depressurization:

- Turn on exhaust fans (bath fans, range hoods, etc), including clothes dryers
- Open/close door to the CAZ (in some houses open will be worst case, in others the door should be closed for worst case)
- Open/close interior doors
- Turn on/off the air handler fan

Table 7. BPI' s retrofit action levels based upon undiluted flue gas CO measurements

CO Level (ppm)	And/Or	Spillage and Draft	Retrofit Action
0-25	And	pass	Proceed
26-100	And	Pass	Recommend fix
26-100	And	Fails (worst case)	Recommend fix
100-400	Or	Fails (natural case)	stop work, fix
>400	And	Pass	stop work, fix
>400	And	Fails (any case)	shut off fuel, fix

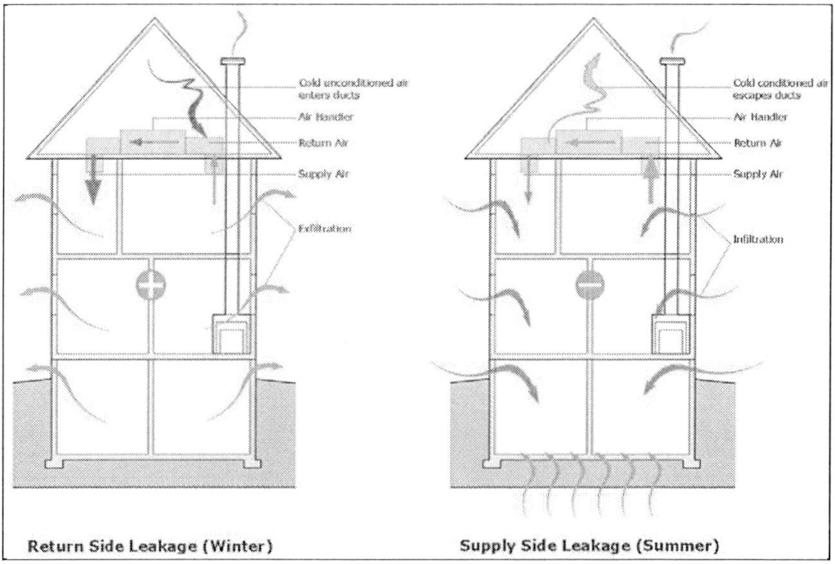

Figure 18. How duct leakage can affect building pressures.

Once you have established the worst-case scenario, turn on the combustion appliances, one at a time, starting with the smallest. Check each appliance individually for spillage and draft. If more than one appliance shares a chimney, also conduct the test with both appliances operating simultaneously.

4.2.5. Implications of Combustion Problems

If combustion problems are identified, they must be addressed before duct sealing. Duct sealing can alter the pressure patterns in a home; different pressure patterns may aggravate combustion problems and increase risks to occupants.

If ducts are located in an unconditioned space like a vented attic, return side leakage can result in unconditioned air and pollutants being pulled into the airstream and can result in the home being positively pressured. This can be an issue during winter months in cold climate regions if moist interior air is pushed into the wall assembly and condenses.

Similarly, attic supply side leakage can result in conditioned air being lost to the outdoors (energy penalty) and the home can become negatively pressurized. If there are atmospheric combustion appliances in the home, these negative pressures may cause back drafting and improper combustion and could result in carbon monoxide poisoning. In hot, humid climate regions, depressurization can also lead to high indoor relative humidity levels as outdoor air infiltration is exacerbated.

Many atmospheric, natural draft, or induced draft appliances require substantial amounts of outdoor air to the CAZ to operate properly. This make-up air or induced infiltration can represent a substantial space conditioning load, and it is not at all desirable in an efficient home. While safe operation can be achieved by providing outdoor air and adjusting appliances properly, installing *sealed-combustion* appliances is very often the best remedy for providing reliable, safe, and efficient space heating and water heating. Strong consideration should be given to this strategy.

In homes with any combustion appliances, carbon monoxide sensors/alarms are required by many codes. These can help alert occupants to serious combustion problems.

Minimum single opening to outside for combustion appliances

Reference: Chapter 24 of 2009 International Residential Code (IRC) for One- and Two-Family Dwellings; Section G2407.6.2

4.3. Personal Safety and Protective Equipment

When working on ducts, recognize that some of the sealant and insulation materials can be hazardous if not handled properly. Refer to the Material Safety Data Sheets (MSDS) for all chemical products used. Use appropriate personal protective equipment including respirators, safety glasses, kneepads, and disposable protective clothing as appropriate.

5. Evaluating Duct Systems

5.1. Check Airflow of System(s)

Furnaces and air conditioners are both designed to run within certain airflow ranges to ensure proper and safe performance. Before sealing ducts, system air flow should be measured. If flow is not in an acceptable range, correct the airflow before continuing with sealing. Sealing a very leaky duct system can increase pressure and therefore reduce total airflow. Air flow should be checked after sealing as well to ensure proper operation.

High airflow during cooling operation may result in inadequate dehumidification. Low airflow is also a problem, as too little heat exchange may occur at the evaporator coil. This can cause liquid refrigerant to return to the compressor and damage the motor. Improper airflow may also cause icing of the evaporator coil.

Below are several methods for measuring airflow through air handlers, any of which may be acceptable if performed properly. Typically, air conditioners require airflow of no less than 300-350 cfm per ton of cooling. Some manufacturers may have somewhat different requirements for particular models, but a 350 cfm/ton minimum may be a good rule of thumb (see side bar for more climate specific "rules of thumb") when the manufacturer's specifications are not readily available. When nominal furnace flow rates are not available during heating operation, 21.7 cfm/kBtu/h is a default value recommended in California energy codes (CEC).

cfm per ton

In humid climates, 350-400 cfm/ton is a common air handler flow range to maximize latent capacity (ability to remove moisture). In drier climates, 400-450 cfm/ton is often used to maximize sensible capacity (ability to lower temperature), as latent capacity is not a significant concern.

5.1.1. Temperature Rise

Furnace airflow may be checked by verifying that the heat rise over the heat exchanger is within the manufacturer's acceptable range. This is done by running the system in heating mode for approximately 5-10 minutes, until the system reaches steady state, while simultaneously taking temperature measurements in the return and supply plenums. The manufacturer's faceplate will have system performance data including the specified heat rise range for that particular model. Typical ranges are 40°F -60°F. Be aware that if the furnace is not operating up to its rated capacity for some reason, this flow verification method may not be accurate.

5.1.2. Duct Pressurization Airflow Testing

The duct pressurization tool ("duct blower") may be used to test for system airflow. This test is done with the registers unsealed, so it should be done prior to setting up duct leakage tests. Refer to the equipment manual for a complete set of instructions on conducting this test. The basic steps are summarized below:

With the system in its normal operating condition, turn the thermostat to the Air Conditioning (AC) fan setting (or on high) and measure the static pressure in the supply plenum.

- Turn the system off.
- Seal off the return side of the system as close to the blower compartment as possible. This is usually done from inside the blower compartment or at the filter slot.
- Once the return has been sealed off, install the duct pressurization fan on the air handler cabinet.
- Monitor the static pressure in the supply plenum and increase the duct blower flow rate until the static pressure under normal operation is achieved. Note: This is the pressure recorded in Step 1.
- Record airflow through the duct blower. This is the total system airflow.

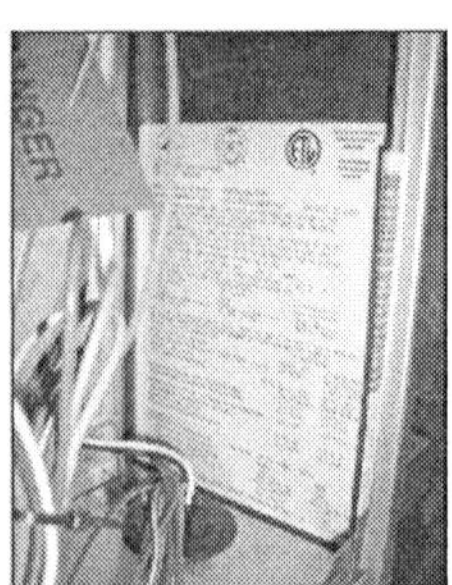

Figure 19. The specified heat rise range for this furnace is 40-70°F according to the faceplate.

Figure 20. Airflow testing using a duct blower.

5.1.3. Flow Plate Airflow Test

A quicker and easier method for testing system airflow is using pitot arrays specially designed for measuring flow in residential air systems. The pitot arrays are integrated into a "flow plate" which should be installed near the blower compartment, typically in the filter slot of the existing system. Velocity pressure readings from the flow plate are easily converted to air flow using manufacturer's instructions.

5.2. Measure Duct Leakage – Duct Pressurization

While it is not necessary to quantify the leakage of a duct system before sealing it, testing can provide insight as to how much improvement is possible (and how much has been achieved after sealing). It is the authors' opinion, however, that when ducts are accessible, they should be sealed. The time and money required for testing may be comparable to effort required to seal the ducts.

The most common duct leakage testing procedures require a variable-speed blower that can be attached to the duct system. The blower includes an orifice and pressure taps with which air flow through the fan assembly can be measured. While equipment manuals—and ASTM E-1554—provide detailed test procedures, the typical procedure is as follows:

- Turn HVAC systems OFF.
- Using tape or other material, block or mask all supply and return registers.
  - Attach the blower to the duct system – typically at the return side of the air handler.
  - Insert a static pressure reference tap—typically in the supply plenum.

- Turn on the fan and pressurize the duct system to the appropriate level (e.g. 25 Pa with respect to outdoors).
- Measure the flow rate through the fan; this is the total duct leakage.

HELPFUL TIP

Discolored duct insulation is an indicator of duct leakage. The insulation is in essence working as a filter of the leaked air.

To more accurately assess energy implications of duct leakage, it is common to quantify the amount of duct leakage escaping "to the outdoors". The test for leakage to the outdoors involves the same duct blower configuration as discussed above, but it also requires a blower door to pressurize the home. Refer to equipment manuals for details, but the basic steps are usually as follows:

- Begin with all HVAC equipment, the duct blower, and the blower door OFF.
- Close all exterior doors and windows in the home.
- With the blower door installed in an exterior door of the home, pressurize the home to the duct test pressure (typically 25 Pa) with respect to outdoors.*

Figure 21. Duct leakage testing.

* It is very important to turn all combustion appliances off before pressurizing or depressurizing the home. If fireplaces or woodstoves have been operated recently, it may not be possible to perform these tests safely. If old ashes are present in a fireplace, lay damp newspaper over the ashes (additional weight may be needed on top) prior to operating the blower door fan.

Figure 22. Blower door being used to pressurize home.

- Turn on the duct blower fan, and increase speed until static pressure in the duct system equals pressure in the home (e.g. 0 Pa wrt indoors, 25 Pa wrt outdoors).
- Measure the flow rate through the duct blower; this is the duct leakage to the *outdoors*.

5.3. Pressure Pan Test

A qualitative test that can sometimes identify leaky duct runs is the "pressure pan" test. This procedure involves using a blower door to depressurize (typically) a home to a reference pressure (e.g. -50 Pa wrt outdoors). When the home is depressurized, a pan is placed over a register, and a manometer measures the pressure differential between the conditioned space and the space just inside the covered register. If the portion of the duct system you're testing is rather well sealed, the differential pressure between the room and the pressure pan should be very small. When large differences are observed, there may be some significant leaks in that part of the system

The pressure pan test is a qualitative test; it can't quantify air leakage. It can be a useful tool to identify the most leaky portions of a duct system.

6. Determine Scope of Work

After conducting preliminary inspections and tests for safety and duct leakage, spend a few moments planning the scope of work for the duct improvement project. Consider:

- Which sections of the duct system can be accessed and effectively sealed/insulated? If there are very few sections of duct accessible, the scope and impact will likely both be small. A method utilizing an aerosol sealant may be worth considering.
- Based on duct leakage tests, what level of benefit can be expected from sealing? If leakage is very low or the forced air system is infrequently used, time and effort may be better spent on other areas of the home's performance.
- Are there major duct disconnects that can be remedied? Re-attaching disconnected ducts will provide huge energy and comfort savings with very little time and effort.
- Are there major holes or gaps that need to be repaired? This may involve new sections of duct and/or the services of a licensed HVAC contractor.
- If sheet metal ducts are insulated, consider the extra effort involved in removing and replacing the insulation; this can make a job more time consuming and challenging.
- Generally, focus sealing efforts on ducts in unconditioned spaces. Reducing duct leakage to conditioned areas may improve comfort somewhat, but it does not typically result in significant energy savings.
- When possible, focus on leaks at and near the air handler. Because system pressures are higher near the fan, benefits from sealing leaks in these areas are greater.

- When atmospheric combustion equipment is located in the same space as the AHU, focus sealing efforts on return ducts to minimize the potential for depressurizing the space.

These are very general guidelines; the exact scope will depend on project goals, time, budget, etc.

Figure 23. Using smoke to visually identify leaks at an air handler.

7. Duct Sealing Procedure

7.1. How to Apply Mastic

The most common complaints about duct mastic are that it is inconvenient or messy to install. With experience, using mastic does not need to be either; sealing with mastic can often be accomplished more quickly and with greater effect than with tapes.

Refer to manufacturer's literature for the proper temperatures to apply mastic; some should not be used in very cold conditions. Before applying mastic, duct surfaces should be relatively clean. While some contractors report that cleaning ducts is less critical for mastics than for tapes, most mastic manufacturer instructions state surfaces must be free of dirt, dust, grease, water, corrosion, etc. While meticulously washing and drying all ducts before sealing may be impractical, some duct systems can be tremendously dusty and dirty. In such cases, it is important to remove much of this grime by wiping or vacuuming.

Mastic is available in pails of various volumes as well as in caulk tubes. Mastic manufacturers usually instruct users to apply mastic with brushes, trowels, or caulk guns. In the authors' experience, applying by hand is often much cleaner, quicker, and more effective. Applying by hand (while wearing appropriate gloves) allows the duct sealer to feel the duct surfaces, seams, and gaps; this is especially important when the gaps are not visible—a very common occurrence.

Using proper gloves is key whether applying by hand or with tools. Be aware of screws and sharp edges. Gloves should protect from some of these risks while keeping hands free from mastic (or at least mostly free from mastic). Care should be taken to remove gloves before touching other surfaces. Once cured, mastic is very difficult to remove from surfaces, clothing, and skin.

Heavy-duty rubber gloves are one option, but these sometimes reduce dexterity and "feeling" capacity when applying mastic by hand. Thin, disposable rubber gloves (e.g. latex gloves), on the other hand, do not offer substantial protection from sharp edges. One compromise involves using inexpensive canvas gloves beneath disposable rubber gloves. The outer latex glove can be replaced when it is too dirty or becomes punctured or torn. The inner glove will last longer, but it can also be discarded when it becomes torn or embedded with mastic.

Another "trick" is to only use one hand to apply mastic. Using one hand—with the appropriate glove(s)—for mastic application will leave the other hand clean to move equipment, reach for tools, crawl through tight spaces, etc. Having one clean hand can keep equipment and job sites much cleaner.

Figure 24. Duct sealer using a single gloved hand (inexpensive cotton glove beneath a disposable latex glove) and a "clean" ungloved hand.

While mastic alone can work well to cover small seams, larger gaps and joints may require mesh tape to reinforce the mastic. Generally, this involves spreading a layer of mastic near the gap, applying mesh tape, and then covering the tape with another layer of mastic. Larger gaps of approximately one inch or more are usually too large for tape and may require patches using sheet metal, duct board, or other approved material.

Follow manufacturer's instructions as to the thickness of applied mastic. If anything, be generous. Some contractors err on the thin side; a very thin coat may not provide adequate sealing or durability (see Figure 25).

Table 8. Duct sealing with mastic and fiberglass mesh tape

1.Spread a layer of masticon the gap.	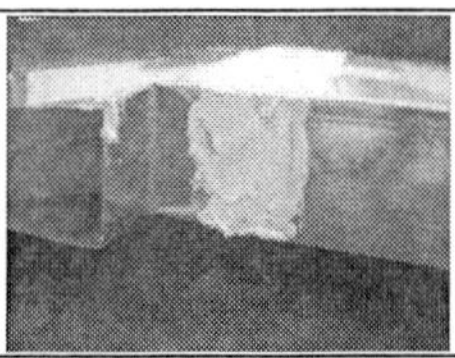
2.Embed the mesh tape intothe mastic.	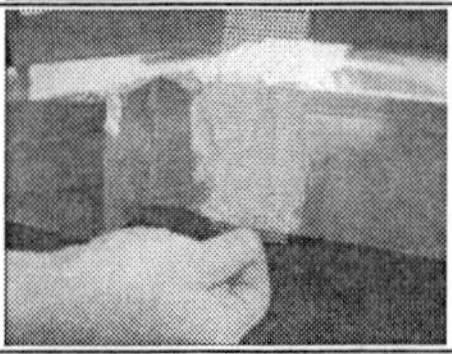
3. Cover the mesh tape with more mastic.	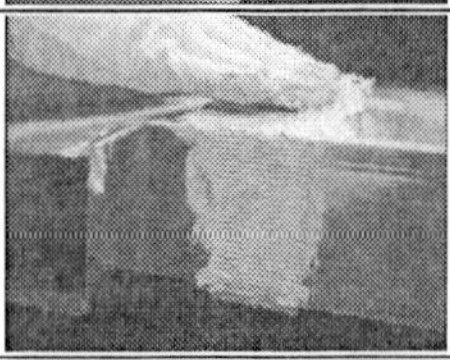
4. Mesh tape should be barely visible.	

7.2. Applying Tape

As with mastics, follow manufacturer instructions for the particular tape used. Duct surfaces should be clean and dry before applying tape. Apply pressure as directed and use at the proper temperatures. Some tapes require heat activation; others require only pressure but do not adhere in cold conditions. Tapes can be especially effective for sealing straight seams and edges where curves do not cause folds or bends in the tape. Tape works well at the transition between the AHU and ductwork, as the tape can be easily cut and replaced during maintenance and replacement of the AHU. Be careful if using your hands to smooth the tape. The edges of foil tape are extremely sharp, a plastic edge or gloves should be utilized.

Figure 25. While sealant has been applied to some of these joints, such a thin coat may not provide an adequate seal. Follow manufacturer's instructions for the thickness when applying mastic.

7.3. Existing Duct Insulation

Sealing sheet metal ducts that have exterior insulation already installed (typically fiberglass wrap) can be much more time consuming than sealing un-insulated ducts. Insulation needs to be removed from accessible portions of the duct system (by carefully cutting insulation, peeling it back, or by removing it completely during sealing). If the insulation is removed completely (with the intent of re-installing after sealing), take care to mark the insulation so it can be reinstalled in the correct position. If the insulation is torn or has a low R-value, it may be worthwhile to install new insulation following sealing.

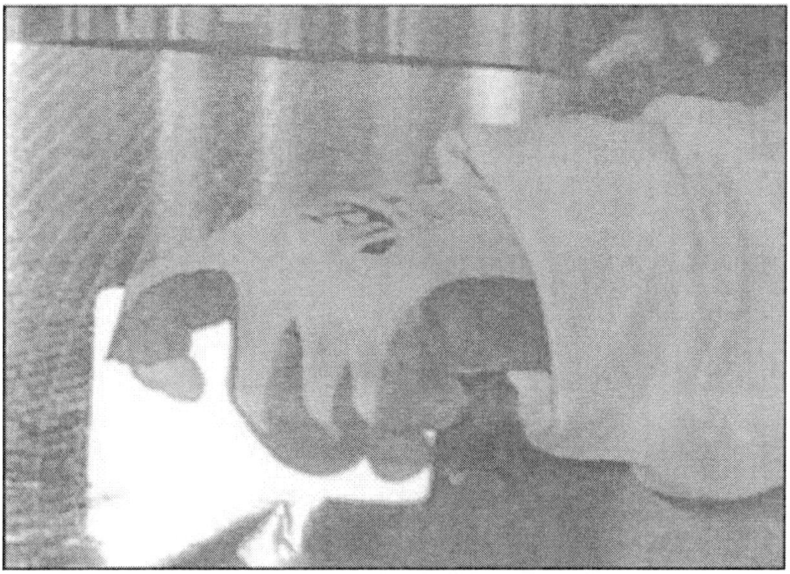

Figure 26. Aluminum tape being smoothed out with a plastic comb.

Be careful if using your hands to smooth foil tape. The edges are extremely sharp

Before removing insulation, take some time to consider how the insulation will be re-installed. In tight spaces, it is much easier to remove insulation than it is to replace it. If testing on a system does not show tremendous leakage, removing insulation, sealing, and re-insulating entire lengths of duct may not be warranted. In some situations it is more practical to focus on other areas (e.g. air handler connections, branch take-offs, boots and registers) where insulation does not need to be completely removed in order to seal penetrations. These areas are sometimes targeted first and followed by another round of testing to determine the need for addressing the other portions of the duct system.

In some older duct systems, ducts may be insulated but lack a vapor barrier. Or, as in the example shown in Figure 28, the vapor barrier has degraded. Insulation air and vapor barriers are important to prevent moisture in hot, humid attics from condensing on cold duct surfaces. Small tears can often be sealed with foil tape; larger gaps may be covered with polyethylene, (foil scrim kraft (FSK), or simply with more duct wrap with an intact air and vapor barrier.

Space and insulation constraints can limit access to leakage areas

Figure 27. Duct sealers here were faced with a tight space and existing insulation. The insulation was cut and temporarily pulled back to seal connections within reach.

Figure 28. Vapor barrier has eroded and needs to be replaced.

7.4. Furnace and Air Handle Connections

The air handler can be one of the leakiest parts of an air distribution system. Because of the higher pressures near the air handler, any holes here have larger impacts than similar-sized holes near registers. For these reasons—and because air handlers are usually somewhat accessible for maintenance—sealing near air handlers should be a primary focus for any duct sealing efforts.

As mentioned above, air handler connections are often good places to seal with tape rather than mastic. When this equipment needs to be serviced and/or replaced, tape can be cut or removed much more easily than can mastic. Air handler connections also tend to be rather straight, with few curves or irregular angles that make sealing with tape more challenging.

A frequent dilemma encountered is how to deal with leakage of the air handler cabinet itself. Many air handlers, especially older units, are responsible for a very large amount of system leakage.[2] Penetrations for wiring, refrigerant lines, condensate drains, gas or water piping, .are often much larger than needed with no gasket or other device to provide air sealing. Sealing such penetrations may limit access to the unit for service. Unfortunately, there is no hard and fast rule for duct sealers other than: seal as much as possible without affecting access, serviceability, or warranty of the unit. Plumber's dope is one potential solution for "non-permanent" sealing of air handler penetrations.

Filter slots are another example of leakage areas that can be tricky to seal effectively. Many newer air handlers have gasketed covers for filter slots. Older handlers typically have an open slot into which a filter is inserted. Taping over the slot can effectively reduce leakage, but this is usually unacceptable because it limits access to the filter. There are some after-market devices that use magnets or removable gaskets to seal these filter slots (see Figure 29).

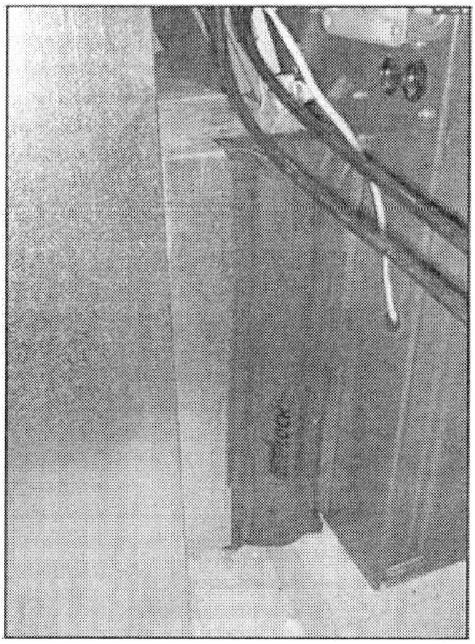

Figure 29. These polymer sheets with magnets at the edges cover the filter slot in an air conditioning system. Covers such as these can reduce leakage tremendously while allowing easy access for filter replacement.

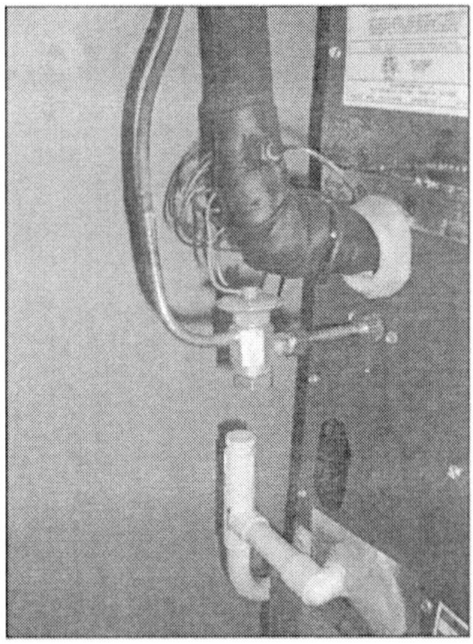

Figure 30. Plumber's putty installed around refrigerant piping to reduce air handler leakage.

Figure 31. Supply and return plenums and trunks sealed with mastic.

7.5. Trunk Ducts

Most residential duct systems consist of one or more main "trunk" ducts from which multiple supply and return lines "branch" off. These trunks are the largest ducts in the home and can be of any material (sheet metal, duct board, or flex – depending largely on the region).

Most duct board trunks are sealed during installation with foil tapes. These tapes do not necessarily maintain their integrity over time, especially if the tapes are relied upon for mechanical support (see Figure 32). Neither tape nor mastic provides mechanical support. As duct board seams are often straight, an appropriate tape may provide adequate sealing. Mastic, however, may prove more durable.

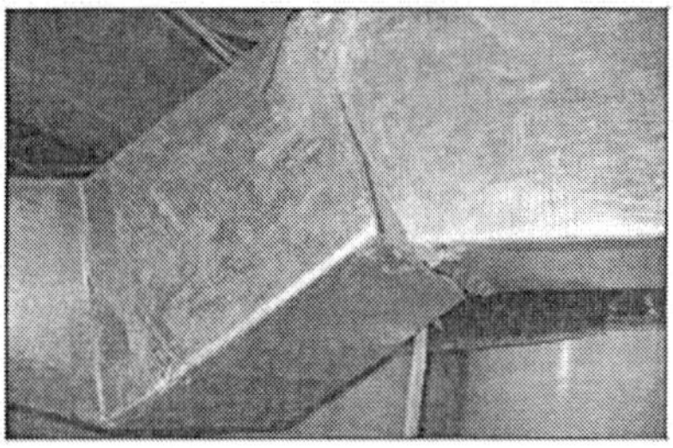

Figure 32. This duct board trunk was sealed with foil tape that has failed due to the stress put on the duct. Neither mastic nor tapes can replace proper mechanical fastening and support.

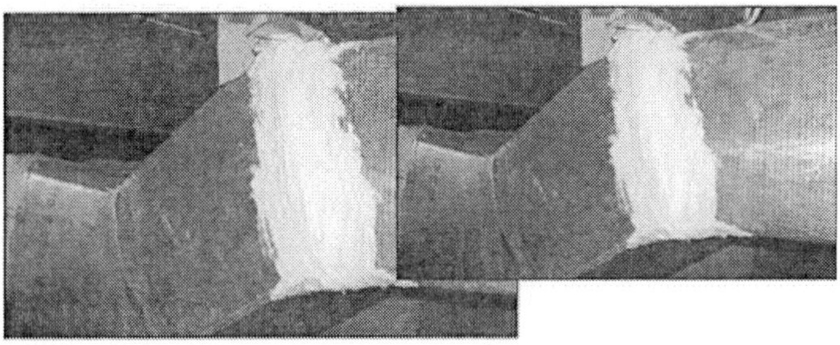

Figure 33. Sealing of duct shown in Figure 32 with mastic and fiberglass mesh.

Sheet metal duct trunks should be sealed with mastic at all joints and seams. Some folded metal duct seams do not leak, but others do. When applying mastic, the time and material needed to cover these seams, as well as joints, is minimal.

7.6. Branch Connections

Branch connections—where smaller supply or return ducts connect with larger trunks—are a key area to target for sealing. Branches are always smaller than trunks, but branch connections are often different shapes (e.g. rectangular trunk to round branch) and different materials (e.g. metal or duct board trunk to flex branch). Because of these variables, many branch connections are sealed poorly, if at all.

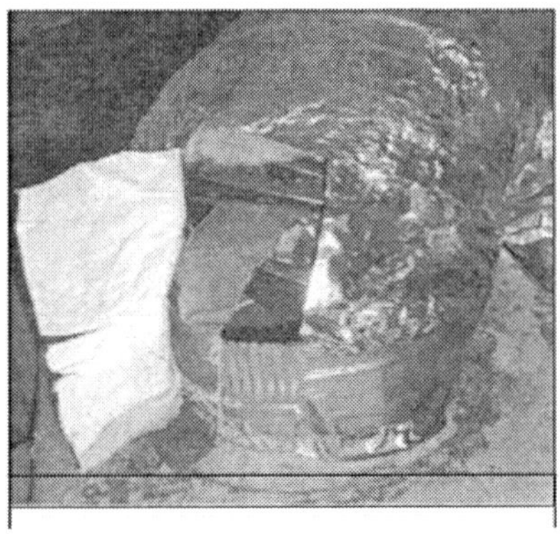

Figure 34. This flex branch never overlapped the collar and the insulation wasn't properly installed resulting in leaks and condensation.

7.6.1. Flex Branches

Insulated, flexible branch ducts have become increasingly common in homes. Occasionally, uninsulated flex duct is used for return branches in conditioned space. Most flex branches connect to trunks via a collar—a round transition piece between the trunk and branch. These collars are connected by various means, but it is common to find little or no sealing of the collars to the trunks. Use mastic to seal collars to the trunks; if gaps are large, mesh tape should be used as well.

With insulated flex, care must be taken to seal all connections with the inner duct lining, not just the outer insulation covering. Sealing the outer insulation layer will not prevent leakage, but will minimize the potential for condensation. It is very common to find insulated flex ducts attached improperly where the outer lining is attached somehow to the collar but the duct itself is not fully connected.

To properly seal flex ducts to metal collars or boots, follow the steps in Table 9. If the existing duct is securely attached to the boot, skip steps 2-4 (step 4 can be done for redundancy).

It is important to seal the branch to the collar connection, as well as the collar to the trunk connection.

Table 9. Properly sealing duct branches to metal collars or boots

1.Ensure that the metal collar is fully accessible (may need to cut back main trunk insulation). Roll the outer insulationlayer back off the inner liner of the flexible duct.	
2.Apply mastic directly to the collar (*best practice, but optional*).	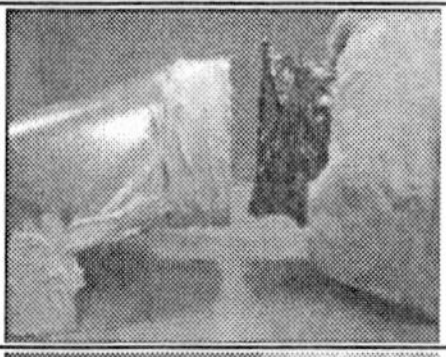
3.Slide the inner flex liner over the collar to embed into mastic.	
4.Secure with a compression band (draw band, zip tie) over the collar.	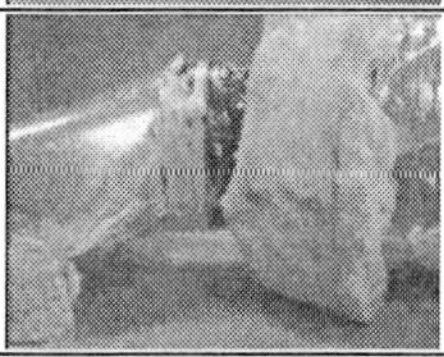
5.Apply additional mastic to seal the seam between the inner liner and collar.	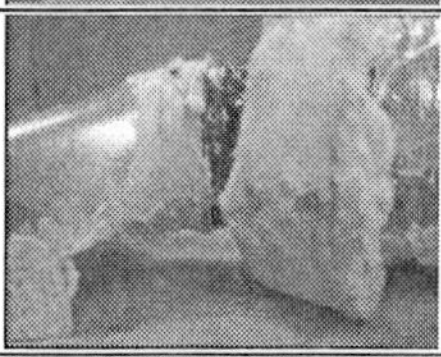
6.Pull the insulation back over the inner liner and secure this joint with another compression band.	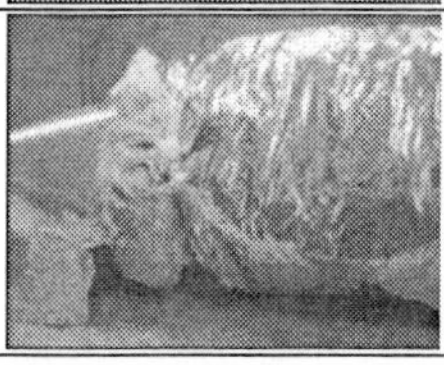

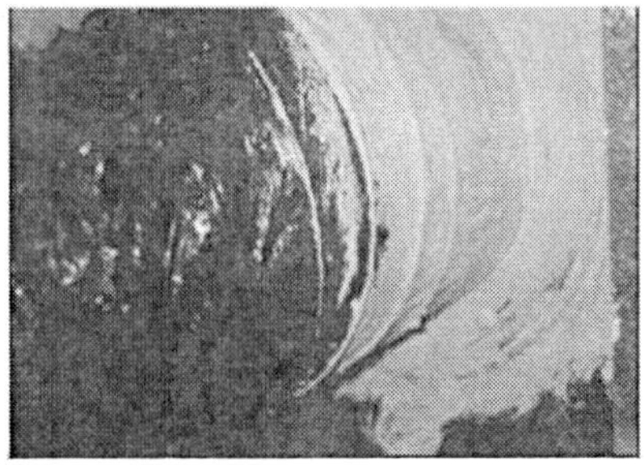

Figure 35. A flex branch (insulation pulled back) attached and sealed to a collar. Note the balancing damper; sealers must be careful that mastic does not prevent access to and operation of such devices.

Figure 36. What exactly is being secured here?

In Figure 40, the inner lining of these flex duct branches have been attached to the collars. However, they have not been sealed. Mastic should be applied where flex duct meets the collar as well as where the collar meets the trunk. If possible, the existing duct insulation should then be stretched to the collar and anchored in place to provide continuous insulation starting at the trunk.

Figure 37. A branch connection collar in a sheet metal trunk. It is important to seal the collar to the trunk as well as the branch to the collar.

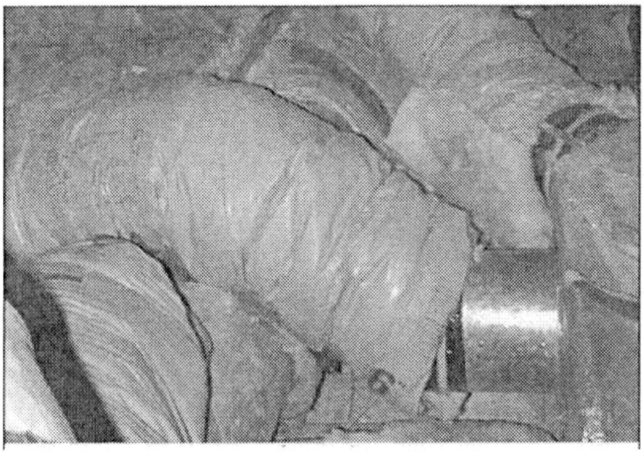

Figure 38. The inner lining of these flex duct branches have been attached to the collars, but they have not been sealed.

Figure 39. Having this much flex duct branching off at the supply plenum makes it very difficult to properly seal the ducts.

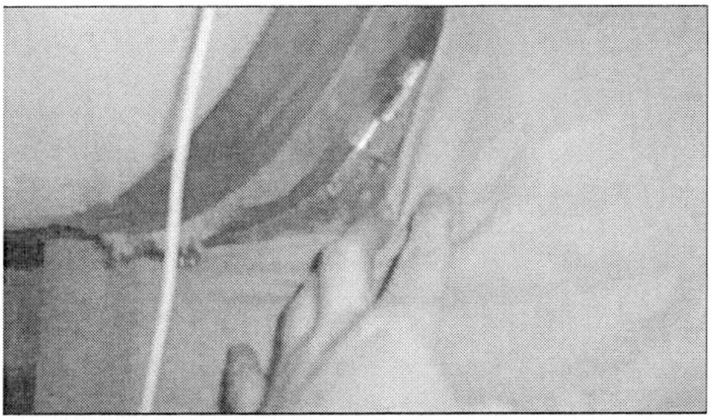

Figure 40. Lack of sealant at the branch connection.

Finding ducts completely disconnected is also not uncommon. For example, in Figure 41, conventional duct tape was the only fastening and sealing method used, and it has failed completely. To rectify the situation, insulation on the trunk should be cut and peeled back so that the collar can be sealed to the trunk with mastic. The inner lining of the branch duct should be attached to the collar (with cable tie or other appropriate means), and this connection should also be sealed with mastic. Insulation should be replaced and taped in place (with the proper tape).

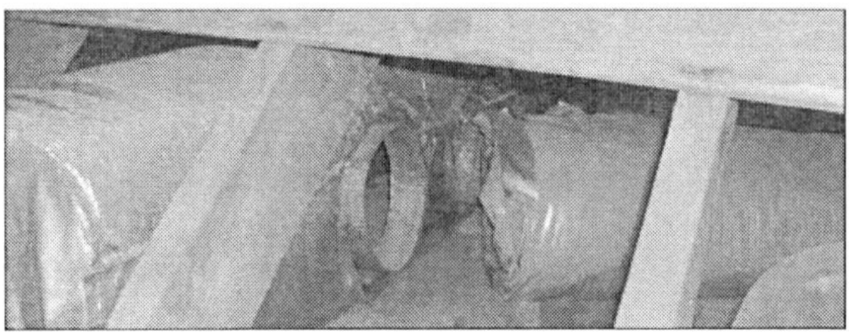

Figure 41. Disconnected ductwork resulting in conditioned air being lost to the vented attic. Homeowner was complaining of comfort issues in a secondary bedroom.

7.6.2. Sheet Metal Branches

For all sheet metal duct connections, any exterior insulation must be peeled back to assess and seal connections. As with flex ducts, sheet metal collars should be sealed to the trunk and branch ducts sealed to the collar. If possible, seal all connections and seams along the entire branch. If the ducts are insulated, time and effort involved in removing (and replacing) duct insulation to access seams may be beyond the scope of the sealing effort.

Figure 42. In existing homes, it may be difficult to see all the duct connections, such as these top take-off sheet metal branches. Use of mastic in these tight situations can be much more effective than tape.

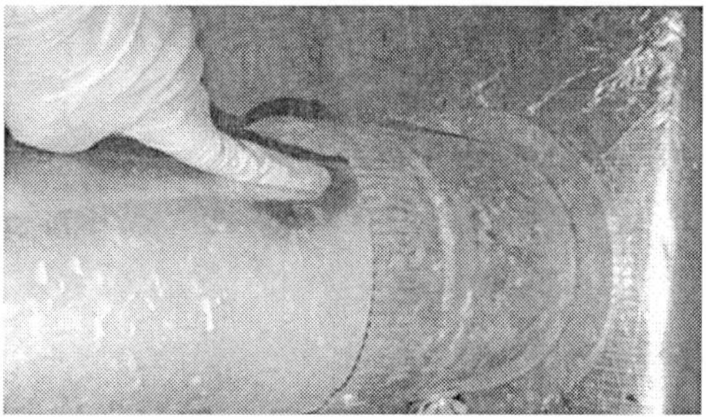

Figure 43. It appears the HVAC contractor ran out of the proper sized ducts or collars for this installation.

Gaps, as shown in Figure 43, result in tremendous energy waste, reduced heating and cooling capacity, and likely comfort problems. The gap is too large to be fixed with mastic or tape; the proper fitting needs to be installed.

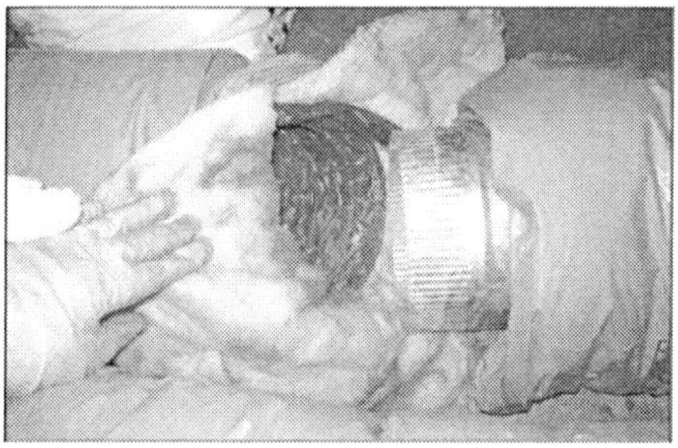

Figure 44. This flex duct junction has failed completely. The inner lining of both sections should be attached and sealed to the metal fitting. Insulation should then be taped in place.

7.7. Branch Duct Fittings

In flex duct branches, there are few connections. A single piece of duct is typically used between the trunk and the duct boot. In longer runs, however, contractors may need to join two sections of flex duct or use duct board junction boxes. These should be sealed much like collars—the inner duct linings should be mechanically attached to the fitting and sealed with mastic.

Sheet metal duct branches have many seams, fittings, and connections. All accessible connections should be sealed with mastic; mesh tape should be used at larger gaps and junctions.

Figure 45. Duct board junction box with multiple flex branch ducts.

Figure 46. A tangle of metal ducts and flue pipes. Branch ducts have been sealed with duct tape which is failing (esp. in lower right). All duct seams and junctions should be sealed with mastic. Care should be taken to avoid contact with the flue piping.

7.8. Boot and Register Connections

As register boots are directly beneath floors, ceilings, or walls, often the easiest way to access them is through the register itself. Removing the grilles from supply and return registers often provides access for sealing:

- Between the branch duct and the boot
- Seams within the boot itself
- Gaps between the boot and the wall, floor, or ceiling surface.

Figure 47. Sheet metal boot that has not been sealed.

Figure 48. Sheet metal boot that has been sealed.

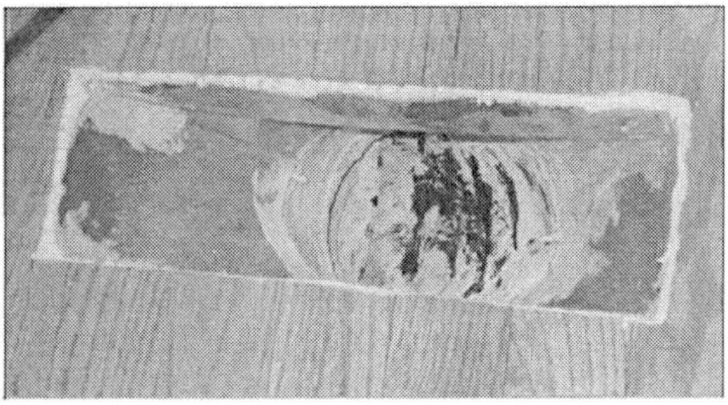

Figure 49. Mastic has been used to seal the flex duct to the register boot, seams in the sheet metal boot itself, and the gap between the boot and the floor. Sealing from the exterior is preferable, but when ducts are not accessible, this method can be utilized. As always, use proper gloves and be careful of screws and sharp edges when sealing boot connections.

Helpful Tip

Make sure not to reinsert the register until the mastic has fully cured to prevent it from being sealed in place.

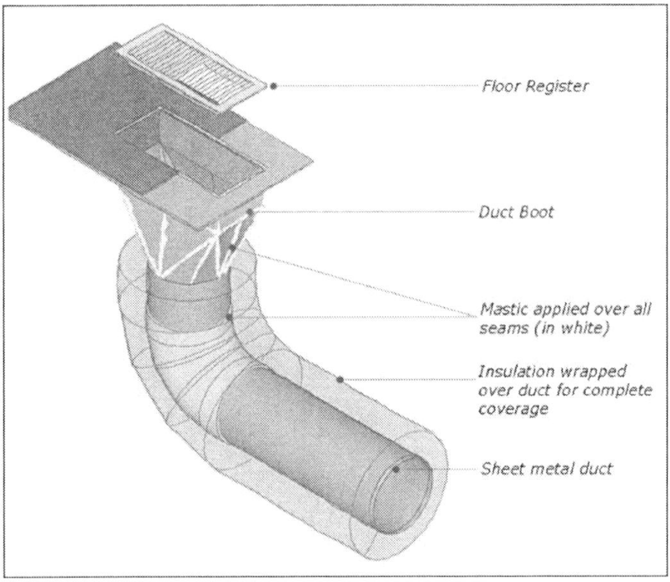

Figure 50. All seams of the sheet metal boot should be sealed with mastic if accessible, either from the interior or exterior of the duct.

These seams should be sealed with mastic, taking care that the mastic does not come in contact with finished surfaces. Often caulk tubes of mastic provide better precision for this type of work.

7.9. Building Cavities

As discussed earlier, using building cavities as part of a duct system is not recommended. One of the key reasons for this is that it is much more difficult to create a tight duct system using building cavities. Nevertheless, panned returns (such as in Figure 51 andFigure 52) are still common, and they are typically very leaky.

Depending upon the scope of the project, it may be preferable for an HVAC contractor to replace panned returns (or other building cavity ducts) with properly sized, sealed, and insulated ducts. If this is not an option, it is possible to improve performance of duct systems by sealing panned returns. Figure 53 shows the area beneath a stairway used as a return plenum. Depressurizing this space drew tremendous amounts of air from the unconditioned basement.

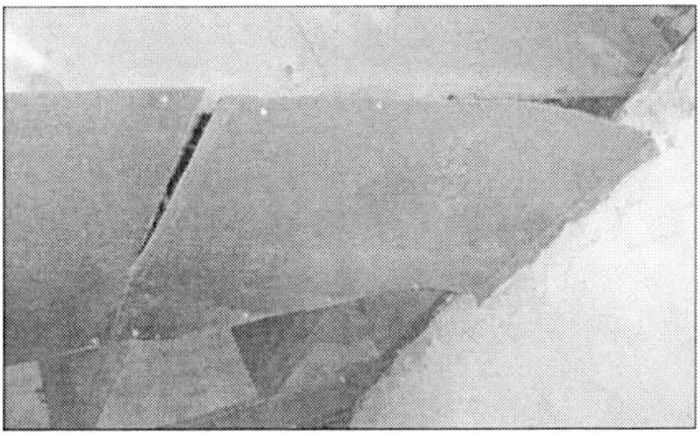

Figure 51. Floor joist bays used as panned returns.

Figure 52. Hard duct, panned returns, and wood blocking are combined as part of the return system. Gaps and seams are not sealed at all, and wiring penetrations allow even more leakage.

8. Insulating Ducts

8.1. Moving Ducts into Conditioned Space

Insulating ducts provides the largest energy benefit when ducts are outside of conditioned spaces (e.g. when they are in unconditioned attics, crawlspaces, basements). Depending on the configuration of the home and on the scope of the project, it may be practical to bring some of these spaces (that contain ducts) inside the envelope, thereby making them conditioned. When ducts are within the conditioned areas, detrimental effects of leakage and conduction are drastically reduced.

Figure 53. This space beneath a stairway was used as a return plenum. Installing a ducted return – the preferred solution – was not in the scope for this project. System performance was improved by sealing this building cavity.

Figure 54. HVAC brought into conditioned space by applying spray foam to roof deck.

Insulating and sealing basement or crawlspace walls, for example, can improve overall performance of the home and HVAC system while making these spaces warmer (in winter), dryer, and more usable. When duct systems are located in an attic, especially in warm climates, air sealing the attic and insulating the roof deck may provide better overall performance than sealing the ceiling plane and improving duct insulation.

Details on these envelope alteration practices are outside the scope of these guidelines. Contractors and homeowners should approach these options holistically, considering the project scope, site-specific details, moisture management, regional codes, etc. More information on these practices are/will be available in these Building America Guidelines:

- Basement Air Sealing and Insulation Methods
- Retrofitting Ducts Into Conditioned Space
- Converting Vented Attic to Unvented.

REMINDER

Even though ducts are moved into conditioned space, an inspection of the duct system is still advised to ensure that there are no major disconnects or other significant faults.

8.2. Insulating Duct Wrap

When sheet metal ducts are located outside of the conditioned spaces in a home, insulation should be added to reduce conductive, convective, and radiant heat gains and losses. Where space and other factors permit, insulation should be installed to at least code-mandated levels (e.g. R-8 for attic ducts and R-6 for ducts elsewhere per 2009 IECC). The most effective way to insulate these ducts is often with "duct wrap" products, typically foil-faced fiberglass blankets.

Follow manufacturer's instructions for installing duct wrap. Most of these instructions begin with language such as, "Ducts shall be clean, dry, and tightly sealed…." This bears repeating: ducts should be thoroughly sealed before insulating. In many situations, reducing duct leakage will have much larger comfort and energy impacts than insulating ducts. Insulating ducts, even with insulation seams tightly sealed, will not reduce duct leakage.

Typical installation instructions for duct wrap follow:

- Use gloves and proper protective equipment.
- Measure the perimeter of the duct to be insulated.

- From manufacturer's literature, determine the proper dimensions of insulation needed to cover duct completely without compression while providing the proper overlap.
- Roll out a section of insulation, measure, and cut appropriately.
- Trim 2" of insulation from all sections of duct wrap that will overlap other sections
- Wrap insulation around the duct; take care that compression is not severe at corners.
- Overlap facing and staple in place per manufacturer instructions.
- After insulation is in place, tape seams with appropriate tape to provide a more rigorous air and/or vapor retarder.

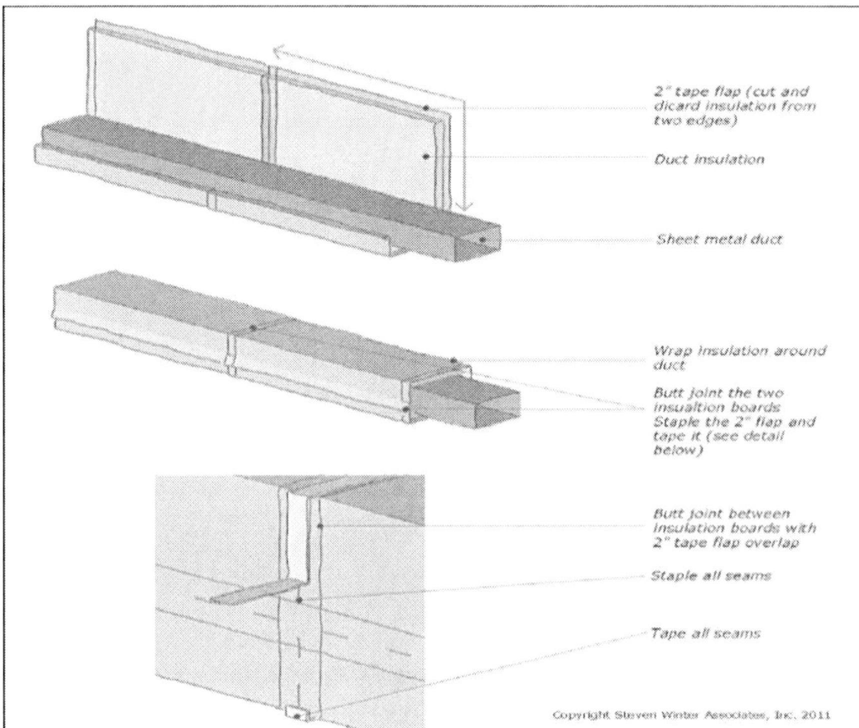

Figure 55. Guidance on properly installing duct wrap.

Of course, in tight spaces with irregularly-shaped ducts, following these instructions may not be straightforward. Duct surfaces should be covered with the target level of insulation to the extent possible. Many ducts are attached directly to floor joists or against walls, preventing insulation from being wrapped completely around them. Insulating ducts in such locations may require some creativity. If ducts are installed flush to floor or ceiling joists, insulation may only be able to cover parts of ducts. Insulation should be sealed as much as possible around these gaps (with appropriate tape, caulks, or mastic).

Table 10. Step-by-step guide to properly installing duct wrap

1. Roll out insulation,measure (add 2” to theduct perimeter), and cutappropriately. Trim 2” ofinsulation from all sectionsof duct wrap that willoverlap other sections.	
2. Wrap insulation aroundthe duct; take care thatcompression is not severeat corners.	 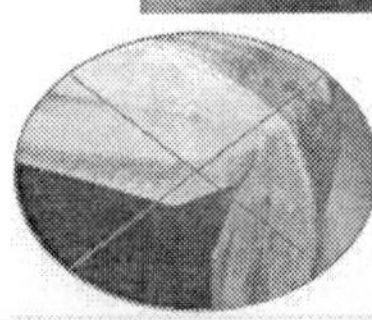 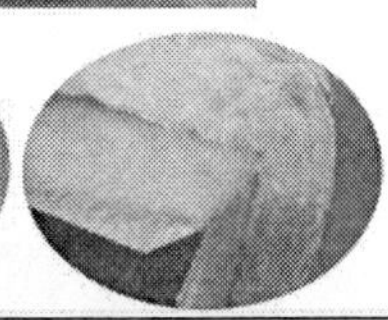
3. Ovcrlap facings	 
4.Staple insulation in place per manufacturer’s instructions. Tape seams with appropriate tape to provide a more rigorous airand/or vapor retarder.	 

There may be areas where rigid insulation (such as fiberglass duct board) may be a more practical material. Duct board may work well in conjunction with flexible duct wrap for parts of sheet metal duct systems. Insulation should be attached and supported mechanically—tapes may not be able to support the weight of the insulation under many conditions.

8.3. Ducts Beneath Attic Insulation

Heating and air conditioning ducts running through 130°F (summer) or 10°F (winter) attics are subject to some pretty hefty thermal penalties. For ducts located in such attics, additional benefits can sometimes be gained by "burying" sealed, insulated ducts beneath spray foam and/or blown attic insulation. This strategy may not be practical in existing homes unless insulating and air sealing of the attic is part of the renovation scope.

Burying ducts beneath attic insulation is only recommended in dry climates. Where attic humidity is high, there is a greater risk that moisture will condense on the cool surface of the duct beneath attic insulation. Ducts can still be buried under the attic insulation but additional steps are required to prevent condensation formation. For more information on this, see Building America Best Practices Series Volume 12: *Builders Challenge Guide to 40% Whole-House Energy Savings in the Cold and Very Cold Climates*.

Figure 56. A duct board trunk partially buried beneath blown attic insulation. This may be impractical in existing homes and in humid climates.

9. Testing Out

Initial testing ("testing in") is important to make sure the home is safe, the systems are in working order, and that sealing will likely not cause any detriment to performance. Testing after the work is complete is also important to verify that systems are in safe, working order after sealing has been performed.

9.1. Duct Leakage Tests

While duct leakage tests are not necessary to assess safety, the tests are useful to gauge impacts of the duct sealing efforts. Duct leakage reductions are required by some programs, and testing duct leakage (both before and after) is one of the best ways for duct sealers to gain experience in the most effective duct sealing practices.

9.2. Airflow

Duct sealing can result in lower air flow across the heat exchangers. Repeat initial tests to see if airflow is still in an acceptable range. If air flow is too low, HVAC contractors may need to reconfigure the system to avoid coil freezing or damaging equipment. Depending on the project, fixes may include increasing the size of ducts to reduce pressure drop; or installing properly-sized, smaller-capacity equipment.

9.3. Combustion

Safety Because duct sealing can change the pressure patterns in a home, it's important to repeat combustion safety tests. If tests fail, the combustion problem must be remedied by a qualified contractor. Remedies may include providing outside air for combustion, or replacing appliances with sealed-combustion.

APPENDIX A. MEASURE IMPLEMENTATION CHECKLIST

The nature and scope of home duct sealing and insulating efforts can vary substantially. Based on these guidelines, the checklist below is presented as a potential framework (though not an exhaustive resource) for conducting duct sealing and insulating efforts.

1. **Inspect The Duct Systems.**

Are the ducts visibly sealed and insulated?	YES	NO
Are they in need of any repairs?	YES	NO

If ducts appear to be well-sealed and insulated, further efforts may not be cost-effective. If the answer is YES for the second question, repairs should be completed before sealing and insulating efforts.

2. **Inspect the home and duct system(s) for significant risks. Are any of the following present?**

Structural concerns	YES	NO
Exposed or knob-and-tube wiring	YES	NO
Severe water or mold problems	YES	NO
Severe insect or rodent infestations	YES	NO
Asbestos (or asbestos-like material)	YES	NO
Other toxic materials	YES	NO
Erratic or dangerous occupants	YES	NO
Occupants with health conditions who may be sensitive to work being done	YES	NO

If the answer is YES to any of these or if other serious risks are identified, do not proceed with work until the risks have been addressed.

3. **System Accessibility**

Are the ducts accessible? Can they be safely accessed to perform testing, sealing, and insulating?	YES	NO

If ducts are not accessible, sealing will likely not be possible with conventional methods.

4. Combustion Safety Testing

4.1. During all tests, measure ambient carbon monoxide (CO) concentrations in the work spaces. Abort work if CO concentrations rise above 35 ppm.

4.2. Identify All Combustion Appliance Zones (CAZ's). for Each Combustion Appliance, Record The Following Information

No.	Location (CAZ)	Make	Model	Fuel
1				
2				
3				

No.	Input Capacity [Btu/h]	Venting Type (e.g. atmospheric, power vent, sealed-combustion, etc.)	Notes
1			
2			
3			

4.3. With all windows and doors closed and appliances turned off, measure the base pressure with respect to outside in each CAZ.

	CAZ	Base Pressure [Pa]

4.4. Establish worst-case depressurization in each CAZ.

- Turn on all exhaust fans and clothes dryer.
- Turn on air handler - if pressure in CAZ decreases (wrt outdoors) then leave air handler on.
- Open and close key interior doors; leave doors in position where CAZ pressure is lowest.

Record the pressure in each CAZ (wrt outdoors) under worst-case conditions.

CAZ	Worst-case Pressure [Pa]

4.5. Fire the smallest combustion appliance in each CAZ. For each appropriate appliance[3] check for spillage, measure draft, and measure CO concentrations. Continue firing appliances until all appliances in each CAZ are operating at once.

No.	Location (CAZ)?	Spillage after one minute	Draft [Pa wrt CAZ]	Flue CO concentration [ppm]
1		YES NO		
2		YES NO		
3		YES NO		

4.6. Continue monitoring ambient CO levels during testing. Record CAZ CO with all combustion appliances.

CAZ	CO concentration with appliances running [ppm]

4.7. Refer to BPI standards (Table 6 in main document) or other guidelines for acceptable test results. If tests fail under worst-case conditions:

- Turn off appliances and allow to cool.
- Establish "natural" conditions (exhaust fans and dryers turned off, interior doors open)
- Repeat the tests (steps 3.4 – 3.5) under "natural" conditions.

Refer to BPI standards for proper action levels. If combustion safety is in question, notify the owner/occupants and do not perform work in the home until the problem is addressed.

5. **Check air handler flow.** Using one of the procedures described in the guidelines, determine the appropriate air flow ranges for each air handler.

Unit No.	Description	Cooling Capacity [tons]	Minimum Cooling Airflow [CFM]	Measured Cooling Airflow [CFM]
1				
2				
3				

Unit No.	Description	Heating Capacity [Btu/h]	Temperature Rise Range [°F]	Acceptable Heating Flow Range [CFM]	Measured Heating Flow [CFM]
1					
2					
3					

6. **Duct Leakage Testing** (optional). If appropriate, conduct leakage tests as described in the guidelines to quantify and identify leaks.

System No	Description	Total Duct Leakage @25Pa [CFM]	Leakage to Outside @25Pa [CFM]	Notes, leakage areas
1				
2				
3				

7. **Develop sealing and insulating plan.** Determine the scope of work; identify accessible ducts with high leakage areas.
8. **Seal ducts.** Using mastic and/or tape as appropriate, seal all accessible ducts. Focus on leaks outside of the conditioned space and near the handler. Record major areas sealed and take pictures as appropriate.

9. **Insulate ducts.** When mastic has set, insulate accessible, uninsulated ducts as appropriate. Focus insulation efforts on ducts outside the conditioned space. Record insulation installed and/or take pictures as appropriate.

10. **Test out.** Repeat combustion safety tests (step 3) and airflow tests (step 4) to assure safe operation. Conduct duct leakage test if appropriate (step 5).

Appendix B. Material Specification

Mastic. Air duct sealant for permanently sealing fabricated joints and seams of HVAC air ducts and thermal insulation.

Required Certification: Water-based duct sealant – UL 181A-M and/or UL 181B-M (depending on duct system type)

Additional Certification: Meets all SMACNA seal and pressure classes.

Relevant Test Methods: ASTM R-96, ASTM D-1310, ASTM E-84, ASTM C-731, ASTM D-2202, ASTM C-732

Aluminum and/or Butyl Tape. Air duct tape for permanently sealing fabricated joints and seams of HVAC air ducts and thermal insulation. All pressure-sensitive tapes used in the manufacture of rigid fiberglass ducts shall be UL 181A listed and UL 181B for flexible ducts.

Required Certification: UL 181A-P, UL 181B-FX

Relevant Test Methods: ASTM D3330, ASTM D882, AAMA 809.2, ASTM E-96, ASTM C711-72, ASTM C763-73

Duct Wrap Insulation. Thermally insulated fiberglass batt material that reduces unwanted heat loss or gain and condensation during system operation.

Required Code Requirements: Meet IRC/IECC insulation requirements for climate.

Relevant Test Methods: ASTM C1290, ASTM C177, ASTM C518, ASTM C976

Closed Cell Spray Polyurethane Foam. Non-structural thermal insulating material that can be spray applied.

Required Code Requirements: IRC - foam plastic, complies with NFPA Standards 286

Relevant Test Methods: ASTM C518, ASTM E84, ASTM E119

Flexible Duct. Thermally insulated, fully lined, flexible air duct designed for use in HVAC distribution systems.

Required Certification: UL 181, complies with NFPA Standards 90A and 90B

Additional Certification: ADC-Certified R-values

Duct Board. Resin-bonded fiberglass formed into rigid rectangular boards faced on exterior with a fire-resistant foil-scrim kraft vapor retarder and faced on the interior with a tightly bonded non-woven mat.

Required Certification: UL 181, complies with NFPA Standards 90A and 90B

Relevant Test Methods: ASTM C 1136, ASTM G21, ASTM G22

Sheet Metal Duct. Fabricated ductwork from galvanized steel sheets. Duct systems shall include UL 181 listed ducts with approved closure systems including collars, connections and splices.

Required Certification: UL 181

Relevant Test Methods: ASTM A653

Drawbands/Compression Band. Adjustable length stainless-steel worm-drive hose clamps or uv-resistant nylon duct ties used to secure flexible ductwork to sheet metal collars. Using an adjustable tensioning tool per manufacturer specifications is recommended.

Required Certification: UL 181A and a minimum tensile strength rating of 50 pounds

References

[1] [2009 IECC] International Code Council. (2009). *International Energy Conservation Code*. Washington, DC: International Code Council.

[2] [2009 IRC] International Code Council. (2009). *International Residential Code*. Washington, DC: International Code Council.

[3] Air Conditioning Contractors of America. (2010). *HVAC Quality Installation Specification*. Arlington, VA: Air Conditioning Contractors of America. https://www.acca.org/Files/?id=693. Accessed April 8, 2011.

[4] [ASTM] American Society for Testing and Materials. (2007). *ASTM E1554 -07, Standard Test Methods for Determining Air Leakage of Air Distribution Systems by Fan Pressurization.* West Conshohocken, PA: American Society for Testing and Materials.

[5] [BPI] Building Performance Institute. (2005*). Technical Standards for the Building Analyst Professional.* Malta, NY: Building Performance Institute. http://www.bpi.org/Web Download/BPI Standards/Building Analyst Professional_2-28-05nNC-newCO.pdf. Accessed April 8, 2011.

[6] Building Technologies Program, Office of Energy Efficiency and Renewable Energy, U.S. Department of Energy. (2003). "*Air Distribution System Installation and Sealing. Building Technologies Program.*" DOE/GO102003-0783, Office of Energy Efficiency and Renewable Energy, U.S. Department of Energy. http://www.toolbase.org/PDF/DesignGuides/doe_airdistributionsysteminstallation.pdf. Accessed April 8, 2011.

[7] [CARB] Consortium for Advanced Residential Buildings. "*Retrofitting America: A 1970s Home Energy Efficiency Analysis*". http://www.carb-swa.com/articles/homepage/RetrofittingAmerica_1970sHomeAnalysis.pdf. Accessed April 8, 2011.

[8] [CEC] California Energy Commission. (2010). 2008 Building Energy Efficiency Standards: Residential Compliance Manual. Sacramento, CA: California Energy Commission. http://www.energy.ca.gov/2008publications/CEC-400-2008-016/CEC-400-2008-016-CMF-REV1.PDF. Accessed September 27, 2011.

[9] Cummings, J. B.; Withers, C. Jr. (January 2008). "Problems Related to Air Handler Leakage." ASHRAE Journal.

[10] [Energy Conservatory] The Energy Conservatory. "*Using a Pressure Pan to Diagnose Duct Leakage*." April 2010. http://www.energyconservatory.com/download/presspanuser.pdf. Accessed September 21, 2011.

[11] [EIA] Energy Information Administration. "Table US7: Unit Price by Fuels Used." 2005 Residential Energy Consumption Survey: Energy

Consumption and Expenditures Tables. http://www.eia.doe.gov/emeu/recs/recs2005/c&e/summary/pdf/tableus7.pdf. Accessed April 6, 2011.

[12] Hohman, J. (2011). *HVACR 301*. Florence, KY: Delmar Cengage Learning.

[13] Holladay, Martin. "*Sealing Ducts: What's Better, Tape or Mastic*?" Green Building Advisor. http://www.greenbuildingadvisor.com/book/export/html/17758 Accessed March 7, 2011.

[14] Moravek, J. (2009). *HVACR 101*. Florence, KY: Delmar Cengage Learning.

[15] Moravek, J.; Hohman J.E. (2010). *HVACR 201*. Florence, KY: Delmar Cengage Learning.

[16] [NREL Duct Insulation] National Renewable Energy Laboratory. "National Residential Efficiency Measures Database: Retrofit Measures for Duct Insulation." National Renewable Energy Laboratory. http://www.nrel.gov/ap/retrofits/measures.cfm?gId=2&ctId=16 Accessed April 7, 2011.

[17] [NREL Duct Sealing] *National Renewable Energy Laboratory. "National Residential Efficiency Measures Database: Retrofit Measures for Duct Sealing.*" National Renewable Energy Laboratory. http://www.nrel.gov/ap/retrofits/measures.cfm?gId=2&ctId=17. (Accessed April 7, 2011).

[18] Rutkowski, H. (2006). *Manual J Residential Load Calculation (Eighth Edition*). Arlington, VA: Air Conditioning Contractors of America.

[19] Rutkowski, H. (2009). *Manual D Residential Duct Systems (Third Edition)*. Arlington, VA: Air Conditioning Contractors of America.

[20] Sherman, M. H., Walker, I. S. & Dickerhoff, D. J. (2000). "*Stopping Duct Quacks: Longevity of Residential Duct Sealants.*" Lawrence Berkeley National Laboratory, LBNL-45423. http://www.osti.gov/bridge/purl.cover.jsp;jsessionid=64967D09B341E3EC57F31C315262B1CB?purl=/776574-DXrFxa/native/. Accessed April 8, 2011.

[21] Silberstein, E. (2004). *Residential Construction Academy: HVACR.* Florence, KY: Delmar Cengage Learning.

[22] [SMACNA] Sheet Metal and Air Conditioning Contractors' National Association. (1998). *Residential Comfort System Installation Standards Manual*. Chantilly, VA: Sheet Metal and Air Conditioning Contractors' National Association, Inc.

[23] [UL 181A] Underwriters Laboratories. (2008). *UL Standard 181A: Standard for Closure Systems for Use With Rigid Air Ducts*. Camas, WA: Underwriters Laboratories.

[24] [UL 181B] Underwriters Laboratories. (2008). *UL Standard 181B: Standard for Closure Systems for Use With Flexible Air Ducts and Air Connectors*. Camas, WA: Underwriters Laboratories.

End Notes

[1] http://www.osha.gov/SLTC/healthguidelines/carbonmonoxide/recognition.html

[2] "Problems Related to Air Handler Leakage." ASHRAE Journal, January 2008 by James B. Cummings and Charles Withers Jr.

[3] Checking for draft and flue CO concentrations is often not appropriate with sealed-combustion, direct-vent, and power-vented appliances. Exhaust CO concentrations can sometimes be measured at the exhaust termination.

In: Guidelines for Improved Duct Design … ISBN 978-1-62081-443-7
Editors: L. M. Collins, D. E. Martinez

Chapter 3

STRATEGY GUIDELINE: ACCURATE HEATING AND COOLING LOAD CALCULATIONS*

Arlan Burdick

FORWARD

The *Strategy Guideline: Accurate Heating and Cooling Load Calculations* report provides information for the following groups:

- Heating Ventilation and Air Conditioning (HVAC) Mechanical Contractors
- HVAC System Designers
- Builders
- House Remodelers.

This guide can be used as a point of entry for collaborative discussions between a builder, HVAC contractor, and other trade partners to understand the importance of appropriately calculating peak heating and cooling loads, as the first step of HVAC system design.

* This is an edited, reformatted and augmented version of Office of Energy Efficiency and Renewable Energy, U.S. Department of Energy, dated June 2011.

Accurate load calculations have a direct impact on energy efficiency, occupant comfort, indoor air quality, and building durability. The load calculation is the first step of the iterative HVAC design procedure, as a full HVAC design involves much more than just the load calculation. The loads modeled by the heating and cooling load calculation process will dictate the equipment selection and duct design to deliver conditioned air to the rooms of the house. This guide references the methodologies of the Air Conditioning Contractors of America (ACCA) publication *Manual J Residential Load Calculation Eighth Edition* (ACCA MJ8), which in turn references information provided by the American Society of Heating, Refrigerating and Air-Conditioning Engineers (ASHRAE). ACCA MJ8 only applies to single family detached dwellings, low-rise condominiums, and townhouses.

This guide is not a new method for performing load calculations, a substitute for established methods of performing load calculations, or step-by-step instructions on how to perform the load calculations. This guide presents the key criteria required to create accurate heating and cooling load calculations and offers examples of the implications when inaccurate adjustments are applied to the HVAC design process. The guide shows, through realistic examples, how various defaults and arbitrary safety factors can lead to significant increases in the load estimate. Emphasis is placed on the risks incurred from inaccurate adjustments or ignoring critical inputs of the load calculation.

Definitions

ACCA	Air Conditioning Contractors of America
ACCA MJ8	Air Conditioning Contractors of America publication *Manual J Residential Load Calculation Eighth Edition*
ACH50	Air Changes per Hour at 50 Pascals
ACHn	Air Changes per Hour Natural
AED	Adequate Exposure Diversity
ALP	Average Load Procedure
ASHRAE	American Society of Heating, Refrigeration and Air-Conditioning Engineers
Btu	British Thermal Unit
Btu/h	Btu per hour
Cfm	Cubic Feet per Minute

°Fdb	Degree Fahrenheit Dry Bulb
°Fwb	Degree Fahrenheit Wet Bulb
HVAC	Heating, Ventilation and Air Conditioning
PLP	Peak Load Procedure
RESNET	Residential Energy Services Network
RH	Relative Humidity
SHGC	Solar Heat Gain Coefficient

EXECUTIVE SUMMARY

The heating and cooling load calculation is the first step of the iterative HVAC design procedure; a full HVAC design involves more than the just the load estimate calculation. Right-sizing the HVAC system, selecting HVAC equipment and designing the air distribution system to meet the accurate predicted heating and cooling loads, begins with an accurate understanding of the heating and cooling loads on a space. The Air Conditioning Contractors of America (ACCA) Manual J Version 8 provides the detailed steps required to calculate the heating and cooling loads. The accurate heating and cooling loads are used to right-size the equipment with ACCA *Manual S Residential Equipment Selection*, then to design the air distribution system and ductwork with ACCA *Manual T Air Distribution Basics for Residential and Small Commercial Buildings* and ACCA *Manual D Residential Duct System Procedure*.

In the author's experience several factors have led to a general industry resistance to initially perform an accurate load calculation, which is necessary for the design of a right-sized HVAC system. Historically, energy codes did not address stringent levels of energy efficiency, and rules of thumb were developed for HVAC sizing that worked based on the construction at that time. Building enclosures have become more energy efficient as energy codes have become more stringent since 2000; however, these rules of thumb have not changed. Full credit should be taken for improvements such as better windows, enhanced air tightness strategies, and additional insulation.

ACCA MJ8 says of safety factors:

> *"Manual J calculations should be aggressive, which means that the designer should take full advantage of legitimate opportunities to minimize the size of estimated loads. In this regard, the practice of manipulating the outdoor design temperature, not taking full credit*

for efficient construction features, ignoring internal and external window shading devices and then applying an arbitrary "safety factor" is indefensible."

"No additional safety factors are required when load estimates are based on accurate information pertaining to the envelope construction and duct system efficiency. Large errors are possible if there is uncertainty about insulation levels, fenestration performance, envelope tightness or the efficiency of the duct runs installed in the unconditioned space."

Examples in this guide show the implications when inaccurate adjustments are applied to the heating and cooling load calculation process. In order to demonstrate the impact on the loads when common "safety factors" are applied to the load calculation process, two houses that meet the 2009 International Energy Conservation Code® (IECC) prescriptive path were modeled as the accurate baseline. One house was modeled in Chicago, IL, and one in Orlando, FL. Several common "safety factors" were applied to the baseline models to create examples of how and where load calculations can be inflated leading to system oversizing. The baseline load calculations were manipulated for:

- Outdoor/Indoor Design Conditions
- Building Components
- Ductwork Conditions
- Ventilation/Infiltration Conditions
- Worst Case Scenario (combining all the safety factors)

Seemingly small manipulations such as changing the outdoor/indoor design conditions can result in exaggerated loads. The Orlando House example (Figure 1) showed a 9,400 Btu/h (45%) increase in the total cooling load, which may increase the system size by 1 ton when the ACCA Manual S procedures are applied.

Combining several adjustments only compounds the inaccuracy of the calculation results. The results of the combined manipulations to outdoor/indoor design conditions, building components, ductwork conditions, and ventilation/infiltration conditions produce significantly oversized calculated loads. The Orlando House example (Figure 2) showed a 33,300 Btu/h (161%) increase in the calculated total cooling load, which may increase the system size by 3 tons (from 2 tons to 5 tons) when the ACCA Manual S procedures are applied. Not only does this oversizing impact the heating and cooling equipment costs, but duct sizes and numbers of runs must also be increased to account for the significantly increased system airflow.

Oversizing the HVAC system is detrimental to energy use, comfort, indoor air quality, building and equipment durability. All of these impacts derive from the fact that the system will be "short cycling" in both heating and cooling modes. To reach peak operational efficiency and effectiveness, a heating and cooling system should run for as long as possible to address the loads. An analogy is that of a car: highway driving at a steady speed will get the best fuel economy, while speeding up and slowing down puts undue wear and tear on the engine and braking systems, and reduces fuel efficiency.

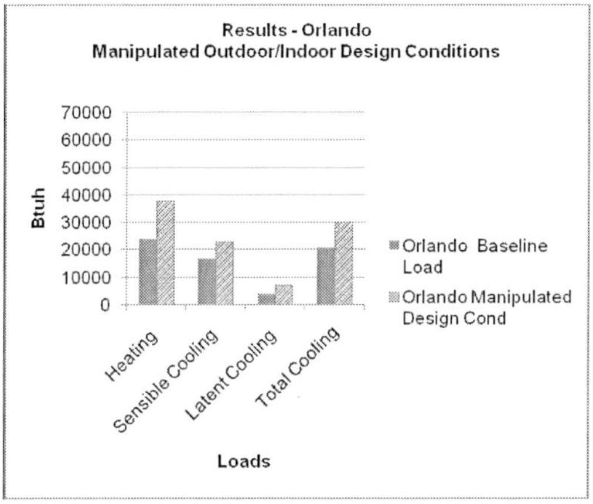

Figure 1. Results – Orlando House Manipulated Outdoor/Indoor Design Conditions.

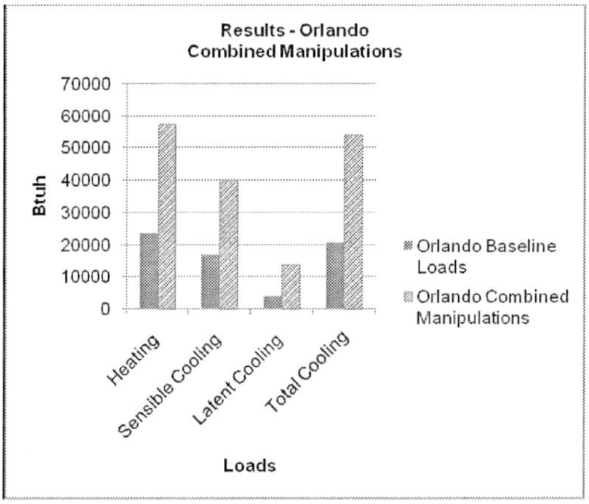

Figure 2. Results – Orlando House Combined Manipulations.

An oversized HVAC system will have both a higher initial cost and a higher cost of operation. The frequent starting and stopping of short cycling can lead to premature failure of the equipment. Short cycling limits the total amount of air circulating through each room, and can lead to rooms that do not receive adequate duration of airflow. In the cooling season in humid climates, cold clammy conditions can occur due to reduced dehumidification caused by the short cycling of the equipment. The system must run long enough for the coil to reach the temperature for condensation to occur and an oversized system that short cycles may not run long enough to sufficiently condense moisture from the air. Excess humidity in the conditioned air delivered to a space may lead to mold growth within the house.

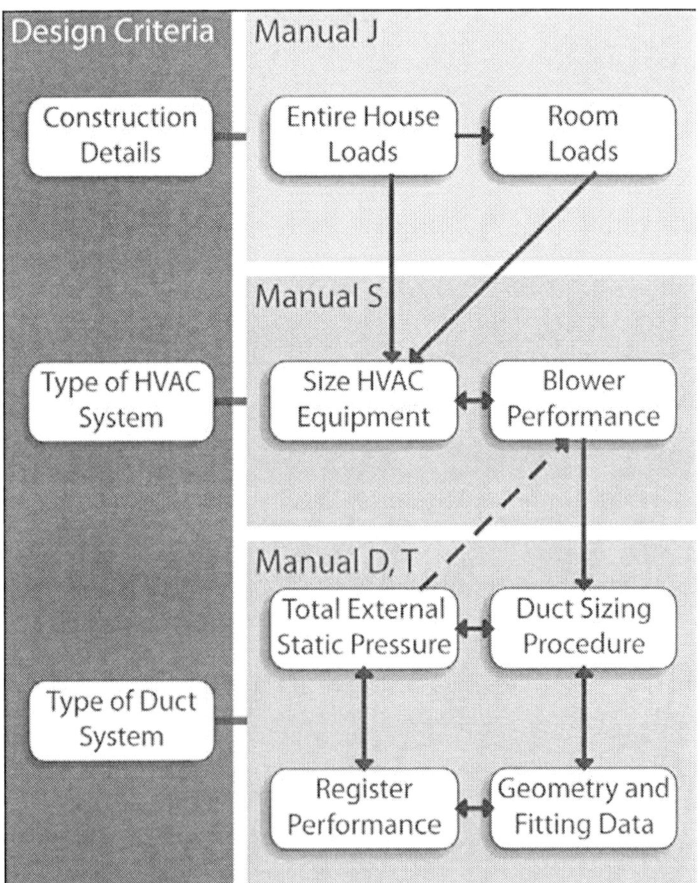

Figure 3. Residential HVAC Design Process.

INTRODUCTION

Heating and cooling loads are the measure of energy needed to be added or removed from a space by the HVAC system to provide the desired level of comfort within a space. Right-sizing the HVAC system begins with an accurate understanding of the heating and cooling loads on a space. Right-sizing is selecting HVAC equipment and designing the air distribution system to meet the accurate predicted heating and cooling loads of the house. The values determined by the heating and cooling load calculation process will dictate the equipment selection and duct design to deliver conditioned air to the rooms of the house, right-sizing the HVAC system. The heating and cooling load calculation results will have a direct impact on first construction costs along with the operating energy efficiency, occupant comfort, indoor air quality, and building durability.

A full HVAC design involves more than the just the load estimate calculation; the load calculation is the first step of the iterative HVAC design procedure (Figure 3). The values calculated from the ACCA MJ8 procedures are then used to select the size of the mechanical equipment. Mechanical equipment selection is done with the aid of the ACCA *Manual S Residential Equipment Selection*. The conditioned air delivery to the space is controlled by the type and size of the air outlet. ACCA *Manual T Air Distribution Basics for Residential and Small Commercial Buildings* provides the guidance on selecting the air outlet size and type. The ductwork that carries the conditioned air to meet the load requirements of the space from the mechanical equipment to the air outlet is sized with the aid of the ACCA *Manual D Residential Duct System Procedure*. The iterative nature of the process involves balancing the blower performance of the mechanical equipment against the pressure losses of the ductwork and air outlets to deliver the conditioned air to the space in as unobtrusive manner as possible. The ACCA procedures have been written into commercial software packages to help the designer work through the many iterations required for a good design.

For the purposes of this guide, an energy-efficient house is defined as one that is designed and built for decreased energy use and improved occupant comfort through higher levels of insulation, more energy-efficient windows, high efficiency space conditioning and water heating equipment, energy-efficient lighting and appliances, reduced air infiltration, and controlled mechanical ventilation. Specification levels for energy-efficient houses have historically been prescribed by beyond code programs that set a percentage better than code for energy use, such as ENERGY STAR® that requires houses to be 15% more energy efficient than code. Beyond-code programs continue to set a percentage better than the improved codes for energy use, raising the bar for whole house energy efficiency. The 2009 International Energy Conservation Code® (IECC) establishes an estimated 15% improvement in energy efficiency over the previous 2006 IECC requirements. Ongoing code cycles are anticipated to incrementally further increase the minimum efficiency of a house. For example, the 2012 IECC achieves approximately 30% savings over the 2006 version. As the new codes are adopted and implemented, a house that was built under an above-code program in 2010 will likely be the code-mandated house in 2015. As the energy efficiency of the house is increased under code or above code programs, the peak heating and cooling loads are significantly reduced.

ACCA MJ8 says of safety factors:

> "*Manual J calculations should be aggressive, which means that the designer should take full advantage of legitimate opportunities to minimize the size of estimated loads. In this regard, the practice of manipulating the outdoor design temperature, not taking full credit for efficient construction features, ignoring internal and external window shading devices and then applying an arbitrary "safety factor" is indefensible.*"

An accurate evaluation of the heating and cooling loads requires a complete understanding and accounting of the building components that make up the thermal enclosure along with the outdoor/indoor contributions to the load. The load calculation for a house where the building enclosure has been enhanced with added air tightness strategies, better windows, and additional insulation will be more sensitive to manipulation of the inputs. In the author's experience, several factors have lead to a general industry resistance to initially perform an accurate load calculation, which is necessary for the design of a right-sized HVAC system. Historically, energy codes did not address stringent levels of energy efficiency, and rules of thumb were developed for HVAC sizing that worked based on the construction at that time. Building enclosures have become more and more energy efficient as energy codes have become more stringent since 2000, but these rules of thumb have not changed. Full credit should be taken by the HVAC system designer for improvements such as better windows, enhanced air tightness strategies, and additional insulation.

HVAC contractors are also not 100% convinced that the builder is managing all other trades to construct a thermal enclosure that truly achieves targeted performance levels. Typically, the HVAC installer will get the comfort complaint if the building enclosure has not met the builder's stated specifications, even though the HVAC system is designed and installed according to the stated specifications. Many HVAC companies rely on the "400 square feet per ton" rule for sizing systems. Even when a load calculation is required to be performed, the contractor may often manipulate the inputs to get a result that is close to the "400 square foot" rule of thumb. Overcoming this bias requires the builder and their trade base sharing an understanding of the importance of achieving the stated building enclosure specifications, and undertaking visual inspections and testing during construction to verify the building meets the intended level of performance. When these checks and balances are in place, the perceived need to add factors of safety during the heating and cooling load calculation process can be avoided. Appendix A contains suggested enhancements to the scope of work for the Mechanical Contractor or HVAC designer.

The critical inputs and their associated risks discussed in this guide are:

- Design Conditions
  - Location
    - Latitude
    - Elevation
    - Outdoor temperature and relative humidity
- Orientation
- Internal conditions
  - Indoor temperature and relative humidity
- Building Enclosure
  - Insulation levels of walls, ceilings, and floors
  - Window specification
    - Thermal conductivity
    - Solar Heat Gain Coefficient (SHGC)
  - Infiltration and ventilation levels
  - Interior and exterior shading
- Internal loads
  - Number of occupants
  - Electronics, lighting and appliances.

A concise checklist to aide in compiling the information required to accurately calculate loads is located in Appendix B of this document.

BASELINE LOAD CALCULATION

In order to demonstrate the impact on the loads when common inaccurate adjustments (also known in the industry as "safety factors") are made to the house, ACCA MJ8 load calculations were performed on two houses (Figure 4) that meet the 2009 International Energy Conservation Code® (IECC) prescriptive path. The first ("Chicago House") is a one story 2,223 ft^2 (above grade) house with a full conditioned basement in Chicago, IL (IECC Climate Zone 5 - CZ5). The other ("Orlando House") is a 2,223 ft^2 one story slab-on-grade house in Orlando, FL (IECC Climate Zone 2 - CZ2). The input conditions used to model the two houses are listed in Table 1.

The calculated loads are listed in Table 2. The peak heating and cooling load is in Btu/h (Btu per hour). The nominal size of the cooling equipment for these two houses is 2 tons, (1 nominal ton = 12,000 Btu/h) based on the calculated cooling loads of 20,600 Btu/h and 20,700 Btu/h for Chicago and Orlando respectively. The Load Dependencies section of this Guide shows the impact on the calculated loads (and therefore the nominal equipment size) when the baseline conditions listed in Table 2 are modified to apply "safety factors" in an effort to come closer to the industry standard sizing of 400 square feet per ton.

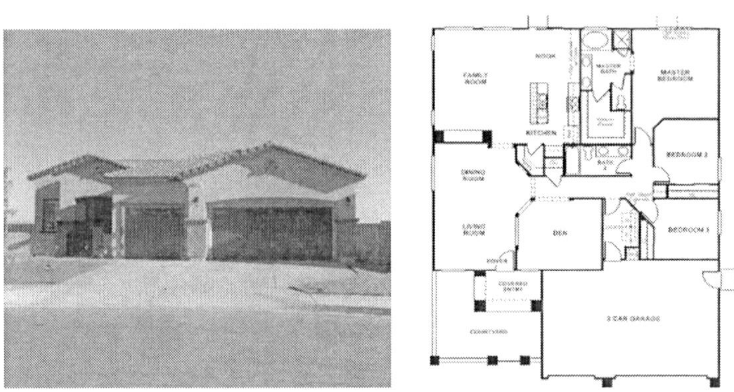

Figure 4. Typical 2223 ft^2 One Story House.

Table 1. Baseline Model Parameters

Condition	Chicago House Climate Zone 5	Orlando House Climate Zone 2
Outdoor Design Conditions	89°Fdb 73°Fwb Temp Cooling 2°F Temp Heating	92°F db 76°Fwb Temp Cooling 42°F Temp Heating
Indoor Design Conditions	75°F Temp 50% RH Cooling 70°F Temp 30% RH Heating	75°F Temp 50% RH Cooling 70°F Temp 30% RH Heating
Roof Assembly	Vented R-38 Attic, Asphalt shingle roofing, 1/2 inch gypsum drywall ceiling	Encapsulated R-31 Attic ("cathedralized attic"), Tile Slate Concrete roofing, 1/2 inch gypsum drywall ceiling
Wall Assembly	9' height, 2x6 Framed walls, R-19, OSB exterior sheathing, Brick exterior, 1/2 inch interior drywall	9' height, Light weight 8" concreteblock 3/4 inch interior XPS, R-4.8, one coat light color stucco, 1/2 inch interior drywall
Foundation Assembly	9' height, 8" poured concrete wall, 2" exterior XPS R10 6.5 feet below grade depth, Heavy damp soil type, concrete slab 8' below grade	Heavy damp soil type, concrete slab-on-grade
Windows	U = 0.35, SHGC = 0.5	U = 0.65, SHGC = 0.30
Ducts	In Conditioned Space	In encapsulated R31 Attic ("cathedralized attic"), Tightness = Supply and Return 0.06 cfm/ft^2, Radial center of room outlets, R-8 insulation
Shading	Full credit for eaves, 50% exterior bug screens, light colored blinds at 45 degrees closed	Full credit for eaves, 50% exterior bug screens, light colored blinds at 45 degrees closed
Infiltration	0.10 ACHn (2.65 ACH50) Cooling 0.19 ACHn (5.03 ACH50) Heating	0.10 ACHn (2.65 ACH50) Cooling 0.19 ACHn (5.03 ACH50) Heating
Ventilation	Balanced supply and exhaust, 60 cfm to meet ASHRAE standard 62.2 with no energy or heat recovery	Balanced supply and exhaust, 60 cfm to meet ASHRAE standard 62.2 with no energy or heat recovery

LOAD COMPONENTS

Heating and cooling load calculations are dependent on the building location, indoor design conditions, orientation, and building construction. The use of commercial software to calculate the heating and cooling loads greatly simplifies the process. Various software applications that provide results certified by ACCA to meet all requirements of MJ8 are available.

Table 2. Baseline Calculated Loads Using ACCA MJ8

Chicago House		Orlando House	
Heating Load	41,700 Btu/h	Heating Load	23,600 Btu/h
Sensible Cooling	17,400 Btu/h	Sensible Cooling	16,600 Btu/h
Latent Cooing	3,200 Btu/h	Latent Cooing	4,000 Btu/h
Total Cooling	20,600 Btu/h	Total Cooling	20,700 Btu/h

Peak Heating Load

The peak heating load represents the amount of heat lost to the outdoor environment at design outdoor and indoor conditions, which must be made up by the HVAC system to maintain occupant comfort (Figure 5). There is one relatively straightforward and uncomplicated heat loss calculation procedure used in ACCA MJ8. The components of the heating load calculation are covered in depth in Section 4 of the ACCA MJ8. The total estimated heat loss is a combination of the sensible heat loss through conduction, infiltration, and ventilation loads. No credit is taken for solar gains or internal loads in calculating the heating load because the peak heat loss occurs at night during periods of occupant inactivity.

Peak Cooling Load

Peak cooling loads represent the amount of heat gained by the house from the outdoor environment at design conditions, which must be removed by the HVAC system to maintain occupant comfort. Cooling loads are made up of the sensible and latent heat gains. The mechanisms of heat gain are conduction, infiltration, ventilation, and radiation (Figure 6). The components of the cooling load calculation are covered in depth in ACCA MJ8.

The cooling load calculation procedure in ACCA MJ8 is more complex than the heating load calculation procedure. ACCA MJ8 documents two cooling load calculation procedures in Section 5-1, the Average Load Procedure (ALP) and the Peak Load Procedure (PLP).

The ALP uses the conditions that are encountered in late afternoon during midsummer when solar gains and temperatures are expected to be highest. This is a simplified method to create dependable cooling load estimates for conventional single family detached houses with a central single zoned comfort system with Adequate Exposure Diversity (AED). See the Orientation section of this Guide for a description of the *Test for Adequate Exposure Diversity*. The ALP procedure minimizes the complexity of the calculations, however, it must be noted the values are averaged for room-to-room loads and approximate the peak block load for the structure.

The PLP looks at time of day sensitivities for midsummer conditions, recognizing that in some designs certain rooms may have significant solar gains during midday or the morning. An example might be a highly glazed sunroom on the south face of a house with considerable east, west and south glass, or a living room with lots of south facing glass and a moderate amount of west glass. The PLP is the residential equivalent to a commercial cooling load procedure, and should be used when a house does not have AED. It is used to estimate peak room loads for the midsummer season and the hour of day in which that room has its maximum solar gain. The whole house cooling load is the sum of the peak for each room.

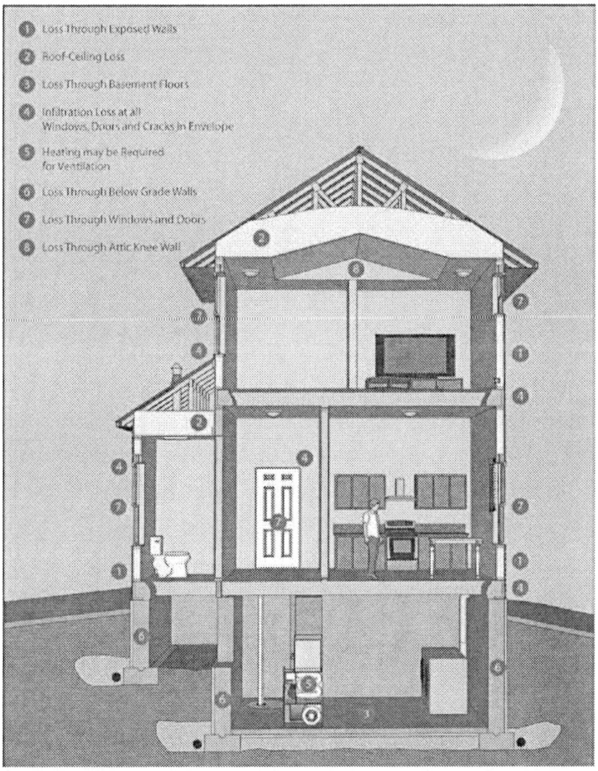

Figure 5. Heat Loss Locations.

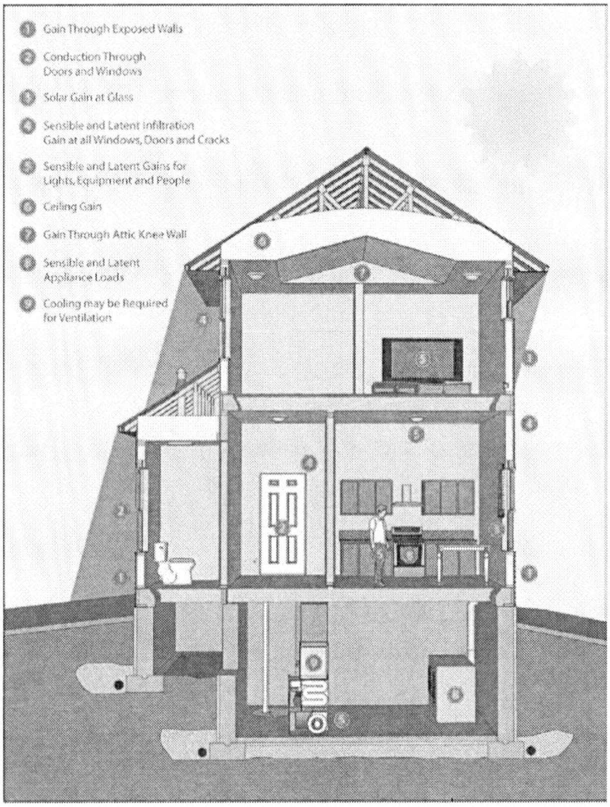

Figure 6. Heat Gain Locations.

Location – Design Conditions

The building location is described by its latitude and elevation. This location describes the values for the outdoor design conditions such as the elevation of the location, latitude, winter heating dry bulb temperatures, summer cooling dry bulb temperatures and relative humidity (RH) information. ACCA MJ8 Table 1A *Outdoor Design Conditions for the United States* has ASHRAE tabulated data for various cities throughout the United States. An example for Orlando, FL, is shown in Table 3. The outdoor design conditions are determined by choosing the city from the table closest to the project city.

Indoor design conditions for the cooling season design conditions of 75°F 50% RH and heating season design conditions of 70°F 30% RH are prescribed in MJ8 and based on the ASHRAE Comfort Zone Chart (Figure 7). These conditions represent the regions of the ASHRAE comfort zone (shaded areas) where, on average, people feel most comfortable. If local codes or regulations specify other design conditions, use those instead of the values in ACCA MJ8 and the ASHRAE tables.

Table 3. Outdoor Design Conditions for the United States-Manual J Version 8, Table 1A

Location		Elevation		Latitude		Winter		Summer	
	Feet	Degrees North	Heating 1% Dry Bulb	Cooling 1% Dry Bulb	Coincident Wet Bulb	Design Grains 55% RH	Design Grains 50% RH	Design Grains 45% RH	Daily Range (DR)
Florida									
Apalachicola	20	29	35	90	78	57	64	70	M
Belle Glade	17	26	44	91	76	42	49	55	M
Cape Kennedy AP	10	28	42	90	78	57	64	70	L
Daytona Beach AP	31	29	37	90	77	51	58	64	L
Fort Lauderdale	10	26	50	90	78	57	64	70	L
Fort Myers AP	15	26	47	93	77	46	53	59	M
Fort Pierce	25	27	42	90	78	57	64	70	M
Gainsville AP	152	29	33	92	77	47	54	60	M
Housestead, AFB	7	25	52	90	76	57	64	70	L
Jacksonville AP	26	30	32	93	77	46	53	59	M
Jacksonville/Cecil Field NAS	80	30	34	95	76	35	43	49	M
Jacksonville, Mayport Naval	16	30	39	92	78	54	61	67	M
Key West AP	4	24	58	89	79	66	73	79	L
Lakeland CO	214	28	41	91	76	42	49	55	M
Melbourne	15	28	43	91	79	62	69	75	M
Miami AP	11	25	50	90	77	51	58	64	L
Miami Beach CO	8	25	48	89	77	52	59	65	L
Miami, New Tamiami AP	10	25	49	91	78	56	59	65	L
Milton, Whiting Field NAS	200	30	31	93	77	46	53	59	M
Ocala	90	29	34	93	77	46	53	59	M
Orlando AP	100	28	42	93	76	39	46	52	M
Panama City, Tyndall AFB	18	30	37	89	79	66	73	79	L
Pensacola CO	30	30	32	92	78	54	61	67	L

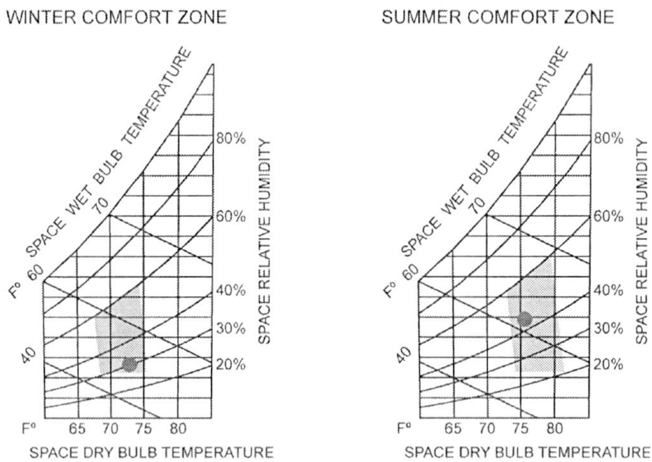

©2009 ASHRAE Handbook Fundamentals. American Society of Heating, Refrigerating and Air-Conditioning Engineers, Inc., www.ashrae.org. Reprinted with permission.

Figure 7. ASHRAE Winter and Summer Comfort Zones.

The outdoor design conditions described by the ASHRAE tables represent long-term average temperatures that will not be exceeded more than a few hours per season. They do not represent the worst weather conditions ever experienced in a given location. Using the Orlando House as an example, Figure 8 shows the temperatures above the outdoor design temperature of 93°F will occur only 13 hours out of the 8,760 hours in a year. The ASHRAE tabulated temperature data is adequate for calculating peak heating and cooling loads, and should not be increased as an additional safety factor.

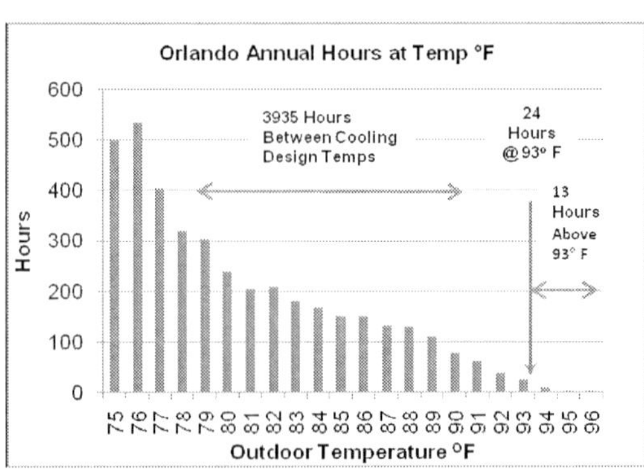

Figure 8. Outdoor cooling bin hours in Orlando, FL.

A right-sized system will run close to 100% of the time at the outdoor design condition and proportionally less at temperatures closer to the inside design temperatures. Designing a system using peak heating and cooling loads based on extreme weather conditions that occur for only a few hours per season, such as the hottest day on record, will result in an oversized system. Over-sized systems in humid climates may not run long enough under some part load conditions to achieve humidity removal from the air and keep the house within the ASHRAE comfort zone.

Orientation

The orientation of the house must be considered in the cooling load calculation due to changing solar heat gains at various times of the day (Figure 9) and the impact of those gains. North, Northeast, East, Southeast, South, Southwest, West, Northwest or North are typically the orientations used for undertaking load calculations for production housing, although the exact cardinal orientation can be used for houses specifically sited on a given lot. The orientation of the house can greatly affect the sensible heat gain on the house depending on the ratio of windows to opaque walls and the degree of shading from the sun.

For production homebuilding, all orientations for given house must be considered. Often times, the peak cooling load for the worst case orientation is acceptable for system sizing; however, if there is a significant difference between loads at various orientations, system sizing may vary for the same house. In the case of the Chicago House, the cooling load varies by 4,400 Btu/h between the worst case and best case orientations of the house (Figure 10). When the HVAC design moves to the system sizing step, Manual S, this variance may impact the system size.

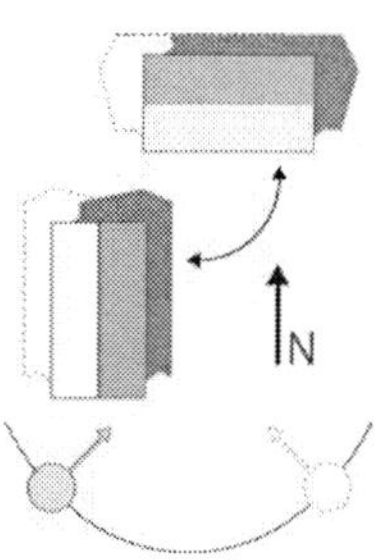

Figure 9. Building Orientation.

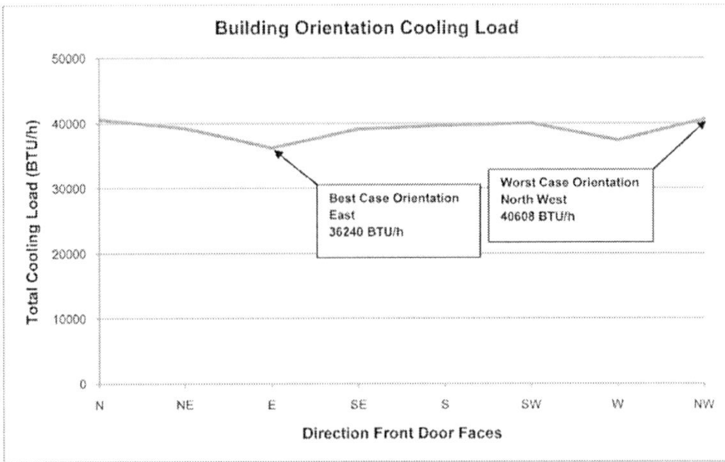

Figure 10. Building Orientation Cooling Load.

The AED limit is an hourly load gain through glazing 30% greater than the average hourly load gain. If a dwelling does not have AED, the architectural features of the dwelling, such as a high percentage of windows in one orientation, will cause a spike in the block cooling load (heat gain) at a particular time of day. The *Test for Adequate Exposure Diversity* is a procedure that is best performed with the aid of software.

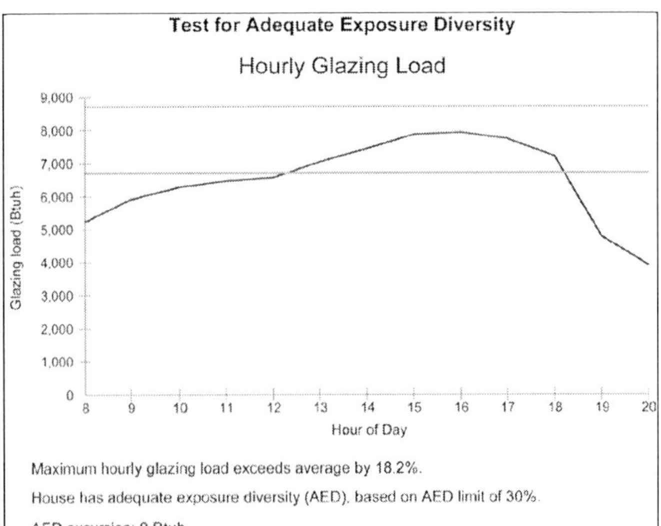

Figure 11. Test for Adequate Exposure Diversity – Pass.

Figure 11 is a software-generated output that shows a house passing the *Test for Adequate Exposure Diversity* by not exceeding that maximum. Figure 12 shows a house that fails by slightly exceeding the AED limit late in the day. The results of the *Test for Adequate Exposure Diversity* require appropriate design judgment in the interpretation of the software output. Appendix 3 of the ACCA MJ8 describes the procedure to compensate for a house that does not pass the *Test for Adequate Exposure Diversity* by compensating for the load only in those rooms with the inadequate exposure diversity without oversizing the whole system. If the variance shown in the test is not addressed, there will likely be a comfort issue in the rooms with the high glazing load.

Building Components

Building construction, proper details, and materials are critical components of the heating and cooling load calculations. The R-value of the building wall, roof, and foundation construction components can be accurately calculated using the insulation levels specified combined with the remainder of the components that make up the construction assembly (i.e. drywall, sheathing, exterior siding materials, structural framing system, roofing materials, etc.). The window performance, described by the U-value and SHGC, must be known and accurately represented by the data input. Shading provided by the overhang of eaves, insect screens, and internal blinds or shades will reduce the sensible heat gain. If shading is ignored in the load calculation the cooling load will be inflated.

Heating and Cooling System Location and Duct Leakage

Best practice for HVAC design is to keep all ductwork within the conditioned space in order to eliminate the duct losses/gains to and from the outside conditions. Scenarios, such as the one-story slab-on-grade Orlando House, present challenges in keeping all ductwork inside conditioned spaces. In a slab-on-grade house, it is typical for an installer to put the HVAC system completely in the attic. Because it has a basement, the Chicago House does not present the same challenges to keeping the ductwork inside conditioned space. In a single-story house with a basement, the duct system is typically run in the basement, which is considered conditioned space provided the basement walls are insulated or there are supply registers in the basement. For ducts outside conditioned space, the heating and cooling loads are more sensitive to duct leakage and R-values of the duct insulation.

Ventilation/Infiltration

Ventilation and infiltration brings outside air into the conditioned space, impacting the heating and cooling load. The target ventilation and infiltration rate must be accurately represented in the data input of the load calculation. In humid climates, the impact on the latent cooling load added by ventilation/infiltration can be significant.

Risks Associated with Oversizing

Oversizing the HVAC system is detrimental to energy use, comfort, indoor air quality, building, and equipment durability. All of these impacts derive from the fact that the system will be "short cycling" in both the heating and cooling modes. To reach peak operational efficiency and effectiveness, a heating and cooling system should run for as long as possible to meet the loads. An analogy is that of a car: highway driving at a steady speed will get the best fuel economy, while speeding up and slowing down puts undue wear and tear on the engine and braking systems, and reduces fuel efficiency.

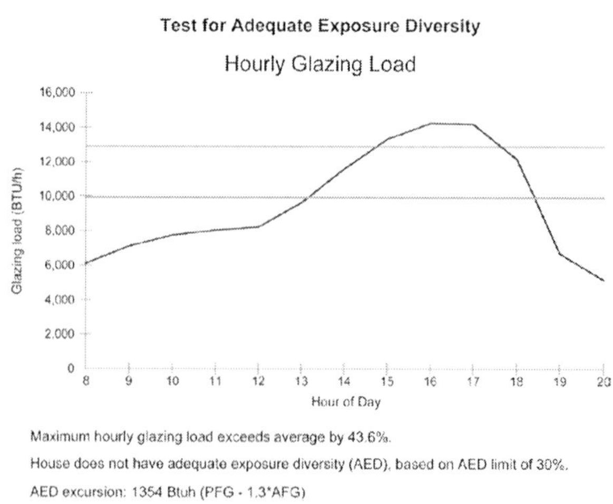

Figure 12. Test for Adequate Exposure Diversity – Fail.

First Cost, Energy Costs and Durability

An oversized HVAC system will have both a higher initial cost and a higher cost of operation. The frequent starting and stopping of short cycling can also lead to premature failure of the equipment.

Comfort - Space Temperatures

Short cycling limits the total amount of air circulating through each room, and can lead to rooms that do not receive adequate duration of airflow. Short cycling of an oversized system can lead to comfort complaints when the spaces located further from the thermostat do not change temperature as quickly as spaces near the thermostat. Even in an energy-efficient house with an enhanced thermal enclosure, this can lead some rooms being colder during the heating season and warmer in the cooling season. In attempt to make the spaces further from the thermostat more comfortable, the occupant may set the thermostat set point higher, requiring additional energy.

Comfort - Humidity Control

The risks associated with oversizing the cooling system, particularly in more humid climates, are also a concern. In the cooling season in humid climates, cold clammy conditions can occur due to reduced dehumidification caused by the short cycling of the equipment. The cooling system removes moisture from the air by passing the air across a condensing coil. The system must run long enough for the coil to reach a temperature where condensation will occur and an oversized system that short cycles may not run long enough to sufficiently condense moisture from the air. Excess humidity in the conditioned air delivered to a space may lead to mold growth within the house.

Load Dependencies

Several common safety factors were applied to the baseline models to create examples of how and where load calculations can be inflated leading to system oversizing.

The baseline load calculations used in this Guide were manipulated for:

- Outdoor/Indoor Design Conditions
- Building Components
- Ductwork Conditions
- Ventilation/Infiltration Conditions
- Worst Case Scenario (combining all safety factors).

Outdoor/Indoor Design Condition Dependencies

To illustrate the effects on the load calculation of manipulating the outdoor/indoor design conditions, the indoor conditions were reversed and the outdoor conditions manipulated to the mean extreme temperatures (Tables 4 and 5). Outdoor temperatures for the manipulated models were taken from the Outdoor Mean Extreme temperatures listed in the ASHRAE 2009 Fundamentals handbook. These changes are representative of those the author has observed from designers wishing to bump the system size up for added cushion.

The results of changing the outdoor/indoor design conditions while keeping all other parameters for the Chicago House the same are shown in Table 6 and Figure 13. When the system is selected, this manipulation would effectively oversize the cooling system by 1 ton.

Table 4. Chicago House Manipulated Outdoor/Indoor Design Conditions

Baseline	Manipulated
Outdoor Design Conditions 89°Fdb 73°Fwb Temp Cooling 2°F Temp Heating	*Outdoor Design Conditions* 97°Fdb 78°Fwb Temp Cooling -11°F Temp Heating
Indoor Design Conditions 75°F Temp 50% RH Cooling 70°F Temp 30% RH Heating	*Indoor Design Conditions* 70°F Temp 30% RH Cooling 75°F Temp 50% RH Heating

Table 5. Orlando House Manipulated Outdoor/Indoor Design Conditions

Baseline	Manipulated
Outdoor Design Conditions 93°Fdb 76°Fwb Temp Cooling 42°F Temp Heating	*Outdoor Design Conditions* 96°Fdb 79°Fwb Temp Cooling 30°F Heating
Indoor Design Conditions 75°F Temp 50%RH Cooling 70°F Temp 30% RH Heating	*Indoor Design Conditions* 70°F Temp 30% RH Cooling 75°F Temp 50%RH Heating

Table 6. Results – Chicago House Manipulated Outdoor/Indoor Design Conditions

	Baseline Load	Manipulated Load	Change In Load Btu/h	Change In Load %
Heating Load	41,700 Btu/h	52,700 Btu/h	11,000 Btu/h	26 %
Sensible Cooling	17,400 Btu/h	24,100 Btu/h	6,700 Btu/h	39 %
Latent Cooing	3,200 Btu/h	6,800 Btu/h	3,600 Btu/h	113 %
Total Cooling	20,600 Btu/h	31,000 Btu/h	10,400 Btu/h	51%

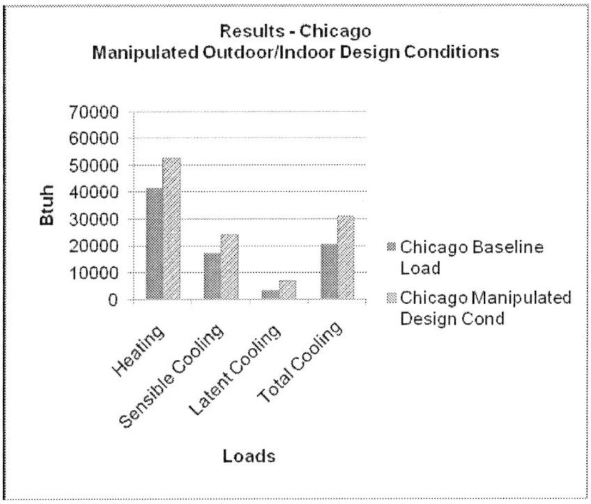

Figure 13. Chicago House Change in Loads, Manipulated Outdoor/Indoor Design Conditions.

The results of changing the outdoor/indoor design conditions while keeping all other parameters for the Orlando House the same are shown in Table 7 and Figure 14. When the system is selected, this manipulation would effectively oversize the cooling system by 1 ton.

Table 7. Results - Orlando House Manipulated Outdoor/Indoor Design Conditions

	Baseline Load	Manipulated Load	Change In Load Btu/h	Change In Load %
Heating Load	23,600 Btu/h	37,800 Btu/h	14,100 Btu/h	60 %
Sensible Cooling	16,600 Btu/h	22,900 Btu/h	6,300 Btu/h	38 %
Latent Cooing	4,100 Btu/h	7,100 Btu/h	3,000 Btu/h	73 %
Total Cooling	20,700 Btu/h	30,100 Btu/h	9,400 Btu/h	45 %

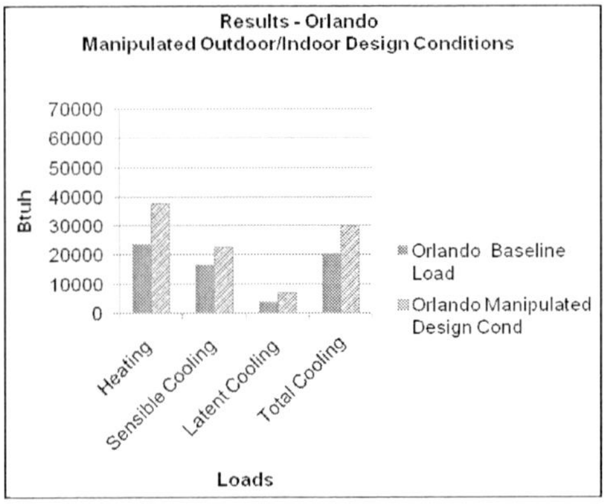

Figure 14. Orlando House Change in Loads, Manipulated Outdoor/Indoor Design Conditions.

Building Components Dependencies

To illustrate the effects on the load calculation of adding a safety factor by manipulating or ignoring known building component values and ignoring shading, the building component specifications were changed as shown in Tables 8 and 9.

Table 8. Chicago House Building Component Manipulations

Baseline	Manipulated
Windows U = 0.35, SHGC = 0.5 *Walls* R-19 *Attic* R-38 Full credit for eaves, 50% exterior bug screens, light colored blinds at 45 degrees closed	*Windows* U = 0.45 SHGC = 0.5 *Walls* R-17 *Attic* R-30 No credit for eaves, no bug screens, no blinds

Table 9. Orlando House Building Component Manipulations

Baseline	Manipulated
Windows U = 0.65, SHGC = 0.30 *Walls* Light weight concrete block w/ .75 inch XPS, R-4.8 *Attic* R-31 sealed Full credit for eaves, 50% exterior bug screens, light colored blinds at 45 degrees closed	*Windows* U = 0.65 SHGC = 0.40 *Walls* Light weight concrete block no XPS R-1 *Attic* R-30 vented No credit for eaves, no bug screens, no blinds

The wall R-value downgrade for the Chicago House is equivalent to the designer attempting to account for a poor installation based on the RESNET Grade 3 installation quality level or ignoring the XPS for the Orlando House. The attic insulation and window downgrade are equivalent to the designer not resetting default values or using values lower that the IECC 2009 levels. These changes are representative of those the author has observed from designers wishing to bump the system size up to for added cushion.

The results of manipulating the building components and shading while keeping all other parameters for the Chicago House the same are shown in Table 10 and Figure 15. When the system is selected, this manipulation would effectively oversize the cooling system by ½ ton.

The results of manipulating the building components and shading while keeping all other parameters for the Orlando House the same are shown in Table 11 and Figure 16. When the system is selected, this manipulation would effectively oversize the cooling system by ½ ton.

Table 10. Results - Chicago House Building Component Manipulations

	Baseline Load	Manipulated Load	Change In Load Btu/h	Change In Load %
Heating Load	41,700 Btu/h	46,300 Btu/h	4,600 Btu/h	11%
Sensible Cooling	17,400 Btu/h	22,400 Btu/h	5,000 Btu/h	28 %
Latent Cooing	3,200 Btu/h	3,200 Btu/h	0 Btu/h	0 %
Total Cooling	20,600 Btu/h	25,700 Btu/h	5,100 Btu/h	24 %

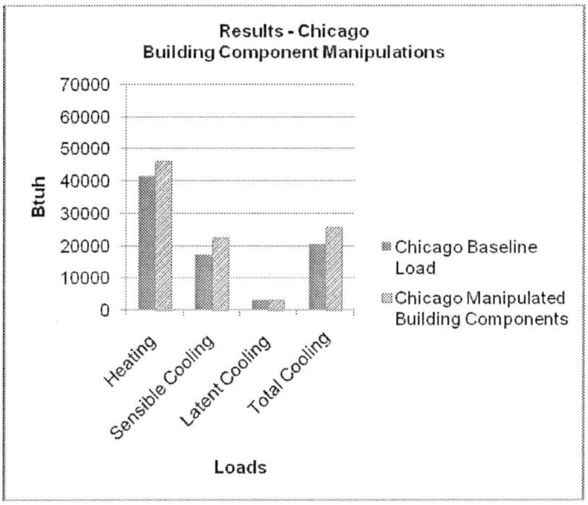

Figure 15. Chicago House Change in Loads, Building Component Manipulations.

Table 11. Results - Orlando House Building Component Manipulations

	Baseline Load	Manipulated Load	Change In Load Btu/h	Change In Load %
Heating Load	23,600 Btu/h	26,600 Btu/h	3,000 Btu/h	12 %
Sensible Cooling	16,600 Btu/h	23,000 Btu/h	6,400 Btu/h	39 %
Latent Cooing	4,100 Btu/h	4,200 Btu/h	100 Btu/h	2 %
Total Cooling	20,700 Btu/h	27,100 Btu/h	6,400 Btu/h	31 %

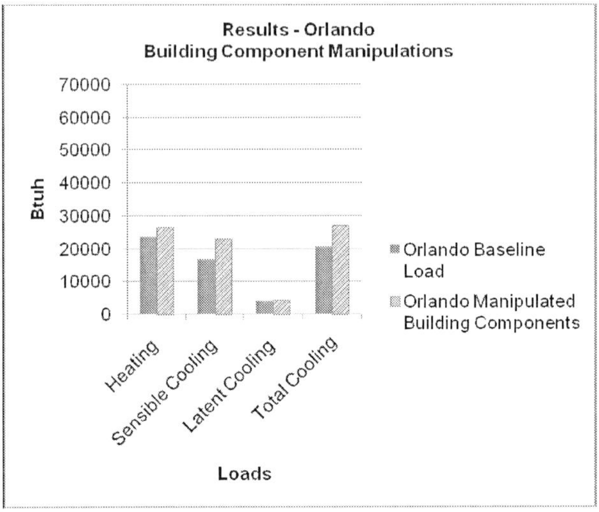

Figure 16. Orlando House Change in Loads, Building Component Manipulations.

Ductwork Conditions Dependencies

To illustrate the effects on the load calculation of adding a safety factor by manipulating or ignoring the ductwork conditions, the specifications were changed as listed in Table 12. This manipulation was not performed on the Chicago House since the Chicago House has an insulated basement and duct work in conditioned space. As discussed in the Heating and Cooling System Location and Duct Leakage section of this Guide, the duct systems in single story houses are typically located in conditioned basements. Basements are considered conditioned space provided the basement walls are insulated or they have supply registers. The manipulated duct tightness values used for the Orlando House are the ACCA MJ8 average duct tightness default values.

Table 12. Orlando House Ductwork Conditions Manipulations

Baseline	Manipulated
Tightness level Supply = 0.06 cfm/ft^2 Tightness level Return = 0.06 cfm/ft^2 Insulation = R-8	Tightness level Supply = 0.12 cfm/ft^2 Tightness level Return = 0.24 cfm/ft^2 Insulation = R-6

The results of manipulating the ductwork characteristics while keeping all other parameters for the Orlando house the same are shown in Table 13 and Figure 17. When the system is selected, this manipulation would effectively oversize the cooling system by ½ ton.

Table 13. Results - Orlando House Ductwork Conditions Manipulations

	Baseline Load	Manipulated Load	Change In Load Btu/h	Change In Load %
Heating Load	23,600 Btu/h	26,700 Btu/h	3,100 Btu/h	13 %
Sensible Cooling	16,600 Btu/h	19,100 Btu/h	2,500 Btu/h	15 %
Latent Cooing	4,100 Btu/h	5,100 Btu/h	1,000 Btu/h	24 %
Total Cooling	20,700 Btu/h	24,200 Btu/h	3,500 Btu/h	17 %

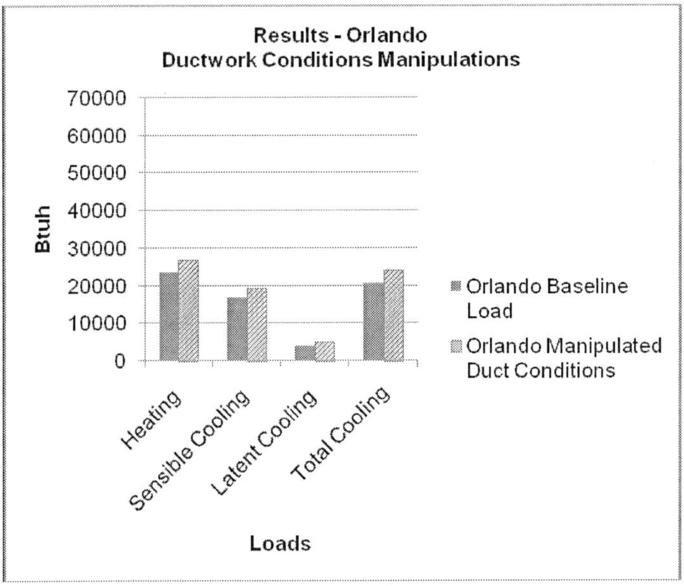

Figure 17. Orlando House Change in Loads, Ductwork Conditions Manipulations.

Ventilation/Infiltration Conditions Dependencies

To illustrate the effects on the load calculation of adding a safety factor by manipulating or ignoring the ventilation/infiltration, Tables 14 and 15 show the specification changes using the ACCA MJ8 default infiltration rate of "semi-loose construction" and a ventilation rate of an arbitrary continuous 100 cfm that may have been determined by looking at the combined effect of simultaneously running all the exhaust fans in the house.

The results of manipulating the ventilation/infiltration rates while keeping all other parameters for the Chicago House the same are shown in Table 16 and Figure 18. This manipulation alone is not enough to increase the cooling system size.

Table 14. Chicago House Ventilation/Infiltration Manipulations

Baseline	Manipulated
Heating season infiltration = 0.19 ACHn (5.04 ACH50) Cooling season infiltration = 0.10 ACHn (2.65 ACH50) Ventilation balanced 60 cfm to meet ASHRAE standard 62.2 without energy or heat recovery	Heating season infiltration = 0.43 ACHn (11.39 ACH50) Cooling season infiltration = 0.23 ACHn (6.09 ACH50) Ventilation exhaust only 100 cfm

Table 15. Orlando House Ventilation/Infiltration Manipulations

Baseline	Manipulated
Heating season infiltration = 0.19 ACH (5.03 ACH50) Cooling season infiltration = 0.10 ACH (2.65 ACH50) Ventilation balanced 60 cfm to meet ASHRAE standard 62.2 without energy or heat recovery	Heating season infiltration = 0.43 ACH (11.39ACH50) Cooling season infiltration = 0.23 ACH (6.09 ACH50) Ventilation exhaust only 100 cfm

Table 16. Results - Chicago House Ventilation / Infiltration

	Baseline Load	Manipulated Load	Change In Load Btu/h	Change In Load %
Heating Load	41,700 Btu/h	46,600 Btu/h	4,900 Btu/h	12 %
Sensible Cooling	17,400 Btu/h	17,900 Btu/h	500 Btu/h	3 %
Latent Cooing	3,200 Btu/h	4,100 Btu/h	900 Btu/h	28 %
Total Cooling	20,600 Btu/h	22,000 Btu/h	1,400 Btu/h	7 %

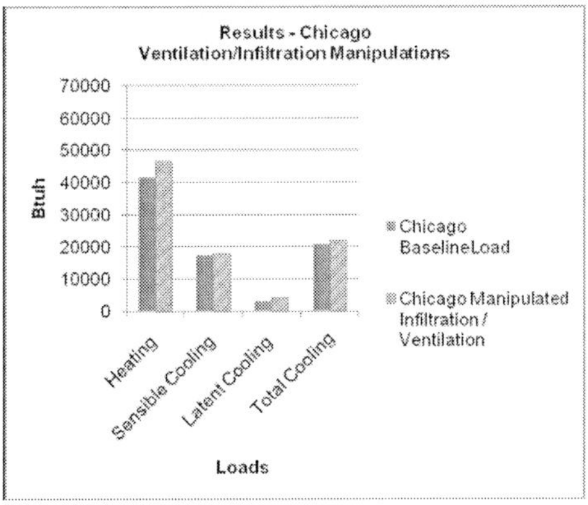

Figure 18. Chicago House Change in Loads, Ventilation/Infiltration Manipulations.

Table 17. Results – Orlando House Ventilation/Infiltration

	Baseline Load	Manipulated Load	Change In Load Btu/h	Change In Load %
Heating Load	23,600 Btu/h	25,600 Btu/h	2,000 Btu/h	9 %
Sensible Cooling	16,600 Btu/h	17,300 Btu/h	700 Btu/h	4 %
Latent Cooing	4,100 Btu/h	5,200 Btu/h	1,100 Btu/h	27 %
Total Cooling	20,700 Btu/h	22,500 Btu/h	1,800 Btu/h	9 %

The results of manipulating the ventilation/infiltration rates while keeping all other parameters for the Orlando House the same are shown in Table 17 and Figure 19. This manipulation alone is not enough to increase the cooling system size.

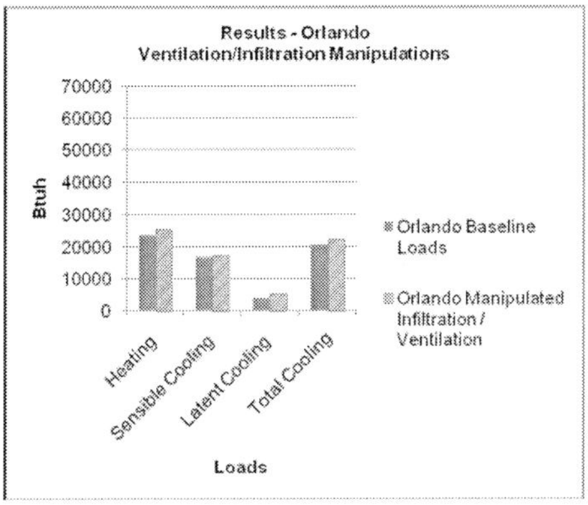

Figure 19. Orlando House Change in Loads, Ventilation/Infiltration Manipulations.

Combined Dependencies

Each of the safety factors applied to the outdoor/indoor design conditions, building components, ductwork conditions, or ventilation/infiltration conditions outlined above has its own impact on the heating and cooling loads. But, a more significant impact occurs when the safety factors are combined. The compounding of safety factors will result in an inflated load that cannot be relied upon for the remaining steps of the HVAC design process. To illustrate the effects on the load calculation of the compounded safety factors, all of the individual impacts were applied to the baseline models.

The results of combining the manipulations while keeping all other parameters for the Chicago House the same are shown in Table 18 and Figure 20. When the system is selected this manipulation would effectively oversize the cooling system by 1½ tons.

Table 18. Results - Chicago House Combined Manipulations

	Baseline Load	Manipulated Load	Change In Load Btu/h	Change In Load %
Heating Load	41,700 Btu/h	64,700 Btu/h	23,000 Btu/h	55 %
Sensible Cooling	17,400 Btu/h	31,600 Btu/h	14,200 Btu/h	82 %
Latent Cooing	3,200 Btu/h	9,100 Btu/h	5,900 Btu/h	184 %
Total Cooling	20,600 Btu/h	40,600 Btu/h	20,000 Btu/h	97 %

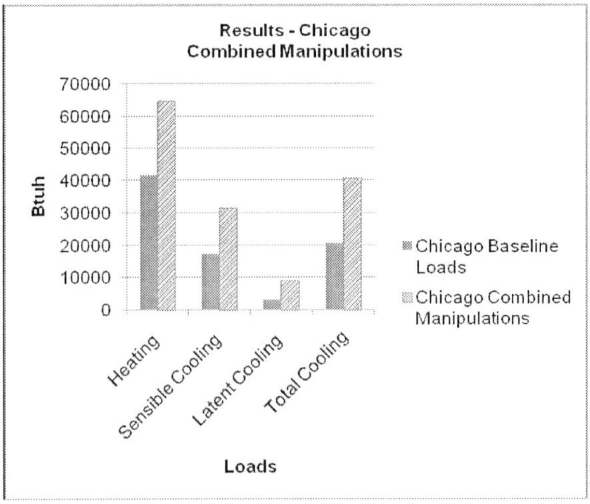

Figure 20. Chicago House Change in Loads, Combined Manipulations.

Table 19. Results - Orlando House Combined Manipulations Results

	Baseline Load	**Manipulated Load**	**Change In Load Btu/h**	**Change In Load %**
Heating Load	23,600 Btu/h	57,200 Btu/h	33,600 Btu/h	142 %
Sensible Cooling	16,600 Btu/h	40,200 Btu/h	23,600 Btu/h	142 %
Latent Cooing	4,100 Btu/h	13,900 Btu/h	9,800 Btu/h	239%
Total Cooling	20,700 Btu/h	54,000 Btu/h	33,300 Btu/h	161 %

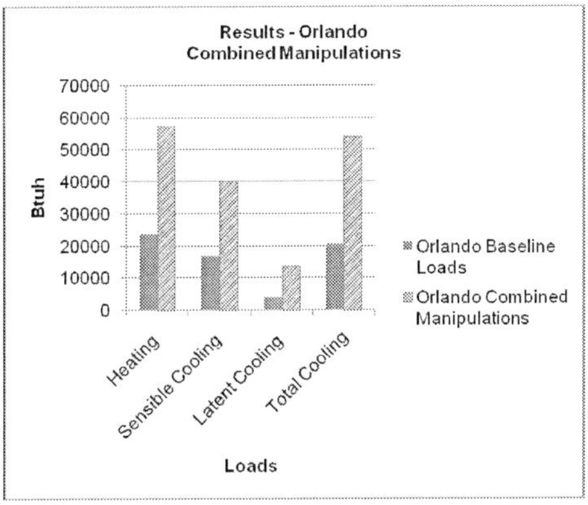

Figure 21. Orlando House Change in Loads, Combined Manipulations.

The results of combining the manipulations while keeping all other parameters for the Orlando House the same are shown in Table 19 and Figure 21. When the system is selected, this manipulation would effectively oversize the cooling system by 3 tons.

CONCLUSION

Right-sizing the HVAC system begins with an accurate understanding of the heating and cooling loads on a space. The values determined by the heating and cooling load calculation process dictate the equipment selection and the duct design needed to deliver conditioned air to the rooms of the house to meet the occupant's comfort expectations.

Examples in this guide showed the implications when inaccurate or inappropriate adjustments are applied during the heating and cooling load calculation process. Seemingly small manipulations such as changing the outdoor/indoor design conditions can result in exaggerated loads. For example, the Orlando House manipulations of outdoor/indoor design conditions showed a 9,400 Btu/h (45%) increase in the total cooling load, which may increase the system size by 1 ton when the ACCA Manual S procedures are applied.

Combining several adjustments only compounds the inaccuracy of the calculation results. The results of the combined manipulations to outdoor/indoor design conditions, building components, ductwork conditions, and ventilation/infiltration conditions produce significantly oversized calculated loads. The Orlando House example showed a 33,300 Btu/h (161%) increase in the calculated total cooling load, which may increase the system size by 3 tons (from 2 tons to 5 tons) when the ACCA Manual S procedures are applied. Not only does this oversizing impact the heating and cooling equipment costs, but duct sizes and numbers of runs must also be increased to account for the significantly increased system airflow.

Again, ACCA MJ8 says of compounding safety factors:

> *"Research studies and the experience of multiple system designers indicate that aggressive use of Manual J procedures provides adequate accuracy. No additional safety factors are required when load estimates are based on accurate information pertaining to the envelope construction and duct system efficiency. Large errors are possible if there is uncertainty about insulation levels, fenestration performance, envelope tightness or the efficiency of the duct runs installed in the unconditioned space."*

The compounding of arbitrary safety factors must be avoided in the load calculation process. The HVAC design industry's practice of using extreme outdoor design conditions, de-rating the insulation value, and over-estimating the infiltration rate render the calculated load values meaningless.

Appendix A: Enhancements to Scope for Mechanical Contractor/ HVAC Manual J Designer

Design Analysis to be completed by Engineer or Contractor, also known as the "Designer"

Calculations to be performed using ACCA Manual J Version 8 Prerequisites:

- All inputs/defaults/building characteristics used for the ACCA Manual J Version 8 calculations must be verified by the HVAC contractor/Mechanical designer in the field and/or must be verified and signed off by the builder that is what is being built consistently across all of the builder's houses in that community or division.
- Design conditions as listed in Table 1A of ACCA Manual J version 8 for the following city shall be used <insert city>. Indoor conditions shall be 75°F Temp 50% RH Cooling and 70°F Temp 30% RH Heating.

The "Designer" shall provide the following information to the builder and HVAC installer:

- Provide room-by-room and building heat gain/loss demand loads based on installed lights and equipment, envelope heat transfer, occupants, required ventilation and all other factors normally considered in these calculations. Loads are to be determined by analysis of actual load calculations performed using ACCA Manual J Version 8 and not by "rules of thumb".
- Determine room supply, return or exhaust air requirements based on the heat gain/loss calculations, total and outside air change requirements, and temperature, humidity, ventilation, and pressurization criteria for each room or space.

Appendix B: ACCA Manual J8 Input Checklist

House Location (city, state) ....________________
House Orientation ....________________

*Complete set of construction drawings including all options are necessary to properly perform a Manual J8 analysis.

Insulation R-Values

Foundation

- Basement Depth below grade ....________________

- Crawlspace Depth below grade ....________________
- Soil Type ....________________
- Wall Type (concrete, ICF, CMU, brick, stone, AWW, metal frame) ....________________
- Foundation Wall Thickness ....________________
- Insulation Type ....________________
- Insulation Location ....________________
- Insulation R-Value ....________________

Walls

Stud Walls

- Stud Spacing ....________________
- Stud Depth ....________________
- Drywall Thickness ....________________
- Exterior Sheathing Type ....________________
- Insulation Type ....________________
- Insulation Location ....________________
- Insulation R-Value ....________________

Non-Stud Walls

- Wall Thickness ....________________
- Drywall Thickness ....________________
- Insulation Type ....________________
- Insulation Location ....________________
- Insulation R-Value ....________________

Floors Over Unconditioned Space

- Sheathing Type ....________________
- Insulation Type ....________________
- Insulation Location ....________________

- Insulation R-Value ....________________

Attic

- Insulation Location ....________________
- Insulation R-Value ....________________
- Roofing Type ....________________
- Sealed or Ventilated ....________________

Windows

- U–value ....________________
- SHGC ....________________
- Low-E or Clear Glazing ....________________
- Number of Panes ....________________

Air Infiltration

- Building Envelope Leakage Rate ....________________

Ductwork

- Insulation Value ....________________
- Location ....________________
- Leakage Rate ....________________

Ventilation

- Ventilation Rate ....________________
- Continuous or Intermittent Ventilation ....________________

Space Conditioning Zoning

- Multi Zone ....________________

- Single Zone ....________________

Stop Work – Do Not Proceed If

IF there is not a complete set of construction drawings clearly identifing areas and dimensions of all exterior surfaces.	**THEN** obtain from the builder a complete set of construction drawings clearly identifing areas and dimensions of all exterior surfaces.
IF the exact specifications for roof/attic assembly i.e. roofing material, radiant barrier, insulation levels are not known.	**THEN** obtain from the builder the exact specifications for roof/attic assembly i.e. roofing material, radiant barrier, insulation levels.
IF the exact specifications for wall assembly i.e. cladding material, insulation levels are not known.	**THEN** obtain from the builder the exact specifications for wall assembly i.e. cladding material, insulation levels.
IF the exact specifications for windows/doors i.e. U-value, SHGC are not known.	**THEN** obtain from the builder the exact specifications for windows/doors i.e. U-value, SHGC.
IF the exact specifications for air tighiness, duct tighness and location, infiltration and ventilation rates are not known.	**THEN** obtain from the builder the exact specifications for air tightness, duct tightness and location, infiltration, and ventilation rates.

References

[1] O'Neal, Dennis L. et.al. ASHRAE 2009 Fundamentals Volume. Atlanta, GA: ASHRAE. 2009.

[2] Rutkowowski PE; Hank. *Manual J Residential Load Calculation Eighth Edition Version Two*. Arlington, VA: ACCA. 2006

[3] National Renewable Energy Laboratory (NREL). Typical Meteorological Year 3 Data Set. NREL Website. http://rredc.nrel.gov/solar/old_data/nsrdb/1991-2005/tmy3/. 2008.

In: Guidelines for Improved Duct Design ... ISBN 978-1-62081-443-7
Editors: L. M. Collins, D. E. Martinez

Chapter 4

STRATEGY GUIDELINE: HVAC EQUIPMENT SIZING*

Arlan Burdick

DEFINITIONS

ACCA	Air Conditioning Contractors of America
AFUE	Annual Fuel Utilization Efficiency
AHRI	Air-Conditioning, Heating, and Refrigeration Institute
ARI	Air-Conditioning and Refrigeration Institute
ASHRAE	American Society of Heating, Refrigerating, and Air Conditioning Engineers
BEopt	Building Energy Optimization (software)
Cfm	Cubic Feet per Minute
EDB	Entering Dry-Bulb
EWB	Entering Wet-Bulb
HSPF	Heating Seasonal Performance Factor
HVAC	Heating, Ventilation, and Air Conditioning
IECC	International Energy Conservation Code
LAT	Temperature of Air Leaving the Cooling Coil
OAT	Outdoor Air Temperature
SEER	Seasonal Energy Efficiency Ratio

* This is an edited, reformatted and augmented version of a Energy Efficiency and Renewable Energy, U.S. Department of Energy, KNDJ-0-40341-02, dated February 2012.

SHR	Sensible Heat Ratio
TD	Room-Air to Supply-Air Temperature Difference

Executive Summary

Right-sizing of a residential heating, ventilation, and air conditioning (HVAC) system involves the selection of equipment and the design of the air distribution system to meet the accurate predicted heating and cooling loads of a house. This Strategy Guideline follows the Air Conditioning Contractors of America (ACCA) *Manual S—Residential Equipment Selection* (Manual S) (Rutkowski 1995) to describe what information is needed and how to use that information to initially select the equipment for a properly designed HVAC system. Equipment selection and duct design are iterative processes where the capacity of the equipment is balanced against the design of the distribution system.

This Strategy Guide describes the equipment selection of a split system air conditioner and furnace for an example house in Chicago, Illinois, as well as a heat pump system for an example house in Orlando, Florida. The required heating and cooling load information for the two example houses was developed by Burdick (2011).

Selection of the equipment can have a substantial impact on the efficiency and operating costs of the system. For the example houses presented in this Strategy Guide, a 92.5% AFUE (annual fuel utilization efficiency) furnace with a 16 SEER (seasonal energy efficiency ratio) air conditioner was the system chosen for the Chicago House and a 9.2 HSPF (heating seasonal performance factor)/16 SEER heat pump for the Orlando House.

With an accurate estimate of the heating and cooling loads completed, the HVAC system designer can begin the equipment selection procedure. It is the system designer's responsibility to select the heating and cooling equipment and to verify that the selected pieces of equipment meet the following:

- The total cooling load, composed of both the sensible load (temperature) and the latent load (humidity)
- The heating load
- The blower capacity (in cubic feet per minute [cfm]) to meet the volume of air cfm range needed.

The general steps to equipment selection are:

1. Determine accurate heating and cooling loads.
2. Select cooling equipment.
3. Select heating equipment.
4. Design the ducts.

It is worth noting that the design of a space conditioning system is an iterative process. Once a duct system configuration and preliminary design are completed, different equipment may need to be selected to achieve the proper airflow to satisfy the room-by-room heating and cooling loads.

1. INTRODUCTION

The heating, ventilation, and air conditioning (HVAC) system is arguably the most complex system installed in a house and is a substantial component of the total house energy use. A right-sized HVAC system will provide the desired occupant comfort and will run efficiently. This Strategy Guideline discusses the information needed to initially select the equipment for a properly designed HVAC system. Right-sizing of an HVAC system involves the selection of equipment and the design of the air distribution system to meet the accurate predicted heating and cooling loads of the house. Right-sizing the HVAC system begins with an accurate understanding of the heating and cooling loads on a space; however, a full HVAC design involves more than just the load estimate calculation—the load calculation is the first step of the iterative HVAC design procedure.

The iterative nature of HVAC design is illustrated in Figure 1, along with the Strategy Guidelines that cover the various aspects of the design process. The heating and cooling loads depend on the building location and construction, whereas the equipment selection and distribution design depend on the loads and each other. After initial equipment selection, several adjustments to duct design and equipment selection may be needed to optimize the design.

The procedures of residential HVAC design are covered in detail by a series of publications produced by the Air Conditioning Contractors of America (ACCA), which in turn references information provided by the American Society of Heating, Refrigerating and Air-Conditioning Engineers

(ASHRAE). The heat loss and gain values are estimated using the procedures from the ACCA *Manual J—Residential Load Calculation* (Manual J) (Rutkowski 2006). Manual J applies only to single-family detached dwellings, low-rise condominiums, and townhouses. Mechanical equipment selection is done with the aid of the ACCA *Manual S—Residential Equipment Selection* (Manual S) (Rutkowski 1995). The ductwork to convey the proper amount of conditioned air to meet the load requirements of the space is designed with the aid of the ACCA *Manual D—Residential Duct System Procedure* (Manual D) (Rutkowski 2009a). Conditioned air delivery to the space is controlled by the type and size of the air outlet. ACCA *Manual T—Air Distribution Basics for Residential and Small Commercial Buildings* (Manual T) (Rutkowski 2009b) provides the guidance on selecting the air outlet size and type. The ACCA procedures have been written into commercial software packages to help the designer work through the iterations required for a good design. Although commercial software is an important tool for design, it should be operated with a solid understanding of the procedures of good design.

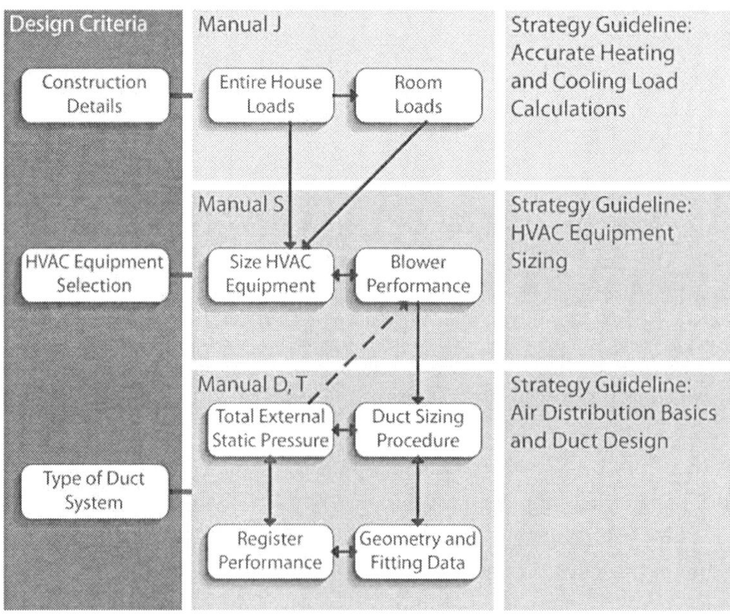

Figure 1. Design information from ACCA Manual J—Residential Load Calculation.

2. Overview

For the purposes of this Strategy Guide, an energy-efficient house is defined as one that is designed and built for decreased energy use and improved comfort through higher levels of insulation, more energy-efficient windows, high-efficiency space conditioning and water heating equipment, energy-efficient lighting and appliances, reduced air infiltration, and controlled mechanical ventilation. Specification levels for energy-efficient houses have historically been prescribed by beyond code programs that set a percentage better than code for energy use. For example, ENERGY STAR® requires houses to be 15% more energy efficient than code. Beyond code programs continue to set a percentage better than the improved codes for energy use, raising the bar for whole-house energy efficiency. The 2009 International Energy Conservation Code® (IECC 2009) establishes an estimated 15% improvement in energy efficiency over the previous 2006 IECC requirements. Ongoing code cycles incrementally increase the minimum efficiency of a house. For example, the 2012 IECC achieves approximately 30% savings over the 2006 version. As the new codes are adopted and implemented, a house that was built under a beyond code program in 2010 will likely be the code mandated house in 2015. As the energy efficiency of the house is increased under code or beyond code programs, the peak heating and cooling loads are significantly reduced.

To illustrate the procedures of HVAC equipment selection, two theoretical houses that meet the 2009 IECC prescriptive path were modeled. The first (the "Chicago House") is a one-story, 2,223-ft^2 (above-grade) house that has a full conditioned basement and is located in Chicago, Illinois (IECC Climate Zone 5 - CZ5). The other (the "Orlando House") is a one-story 2,223-ft^2 slab-on-grade house in Orlando, Florida (IECC Climate Zone 2 - CZ2).

In addition to the climatic differences between the two house locations, the Chicago House is a house over a full basement, whereas the Orlando House is a slab-on-grade foundation, allowing for the discussion of different mechanical and ductwork layouts.

Figure 2 shows the Chicago House example split system with an upflow furnace and evaporator coil located in the basement and an outdoor condenser coil. Figure 3 shows the Orlando House example air source heat pump with outdoor coil and horizontal fan coil in the conditioned attic.

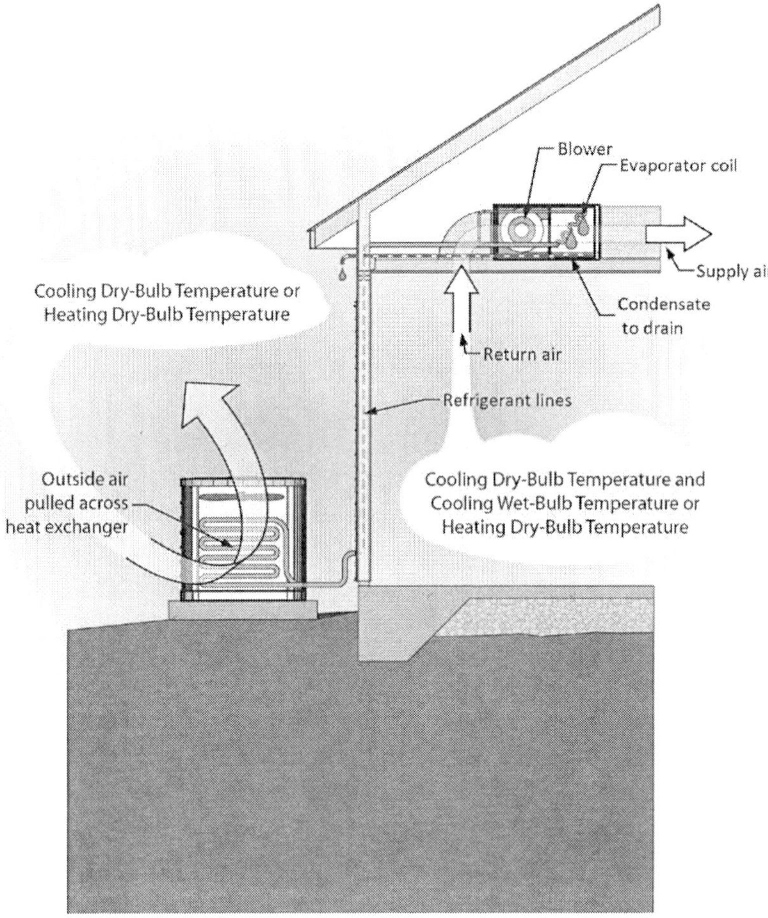

Figure 2. Upflow furnace scenario as in the Chicago House.

2.1. Design Conditions

The impacts and importance of accurate calculation of heating and cooling loads are covered in depth by Burdick (2011). The equipment selection discussion in this Strategy Guide builds upon the information developed in that heating and cooling loads guide. The Manual J loads from Burdick (2011) are listed in Table 1.

Table 1. ACCA Manual J Heating and Cooling Loads

Chicago House		Orlando House	
Heating Load	41,660 Btu/h	Heating Load	23,640 Btu/h
Sensible Cooling	17,370 Btu/h	Sensible Cooling	16,630 Btu/h
Latent Cooling	3,240 Btu/h	Latent Cooling	4,060 Btu/h
Total Cooling	20,610 Btu/h	Total Cooling	20,690 Btu/h

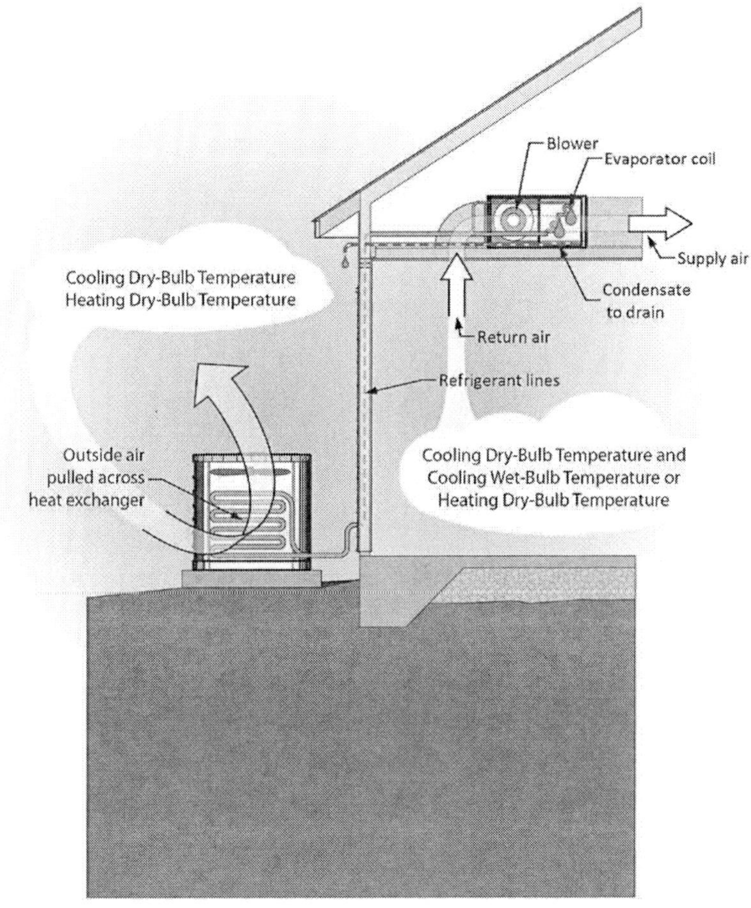

Figure 3. Heat pump scenario as in the Orlando House.

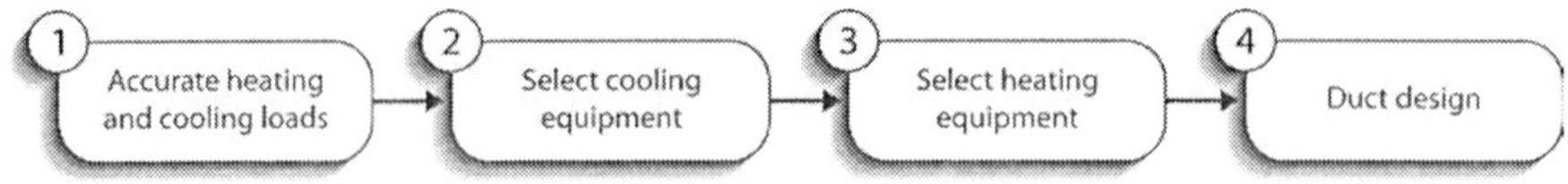

Figure 4. General steps to equipment selection.

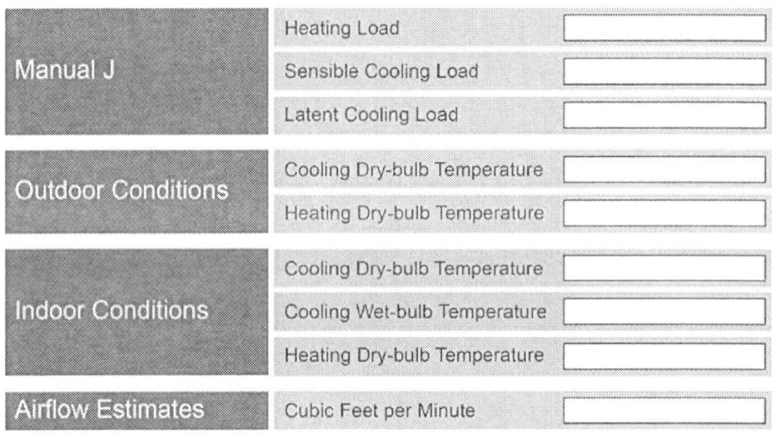

Figure 5. Design information needed.

The general steps to equipment selection are shown in Figure 4, and the design information needed to select equipment is outlined in Figure 5.

Note that the design of a space conditioning system is an iterative process, and once a duct system configuration and preliminary design are completed, different equipment may need to be selected to achieve the proper airflow to satisfy the room-by-room heating and cooling loads.

The design loads along with the indoor and outdoor conditions information needed for equipment selection are all taken from the inputs used in the Manual J procedure. Faithful adherence to the conditions used in Manual J is required for accurate equipment selection.

2.2. Preliminary Equipment Selection

With an accurate estimate of the heating and cooling loads completed, the designer can begin the equipment selection procedure. It is the system designer's responsibility to select the heating and cooling equipment and to verify that the selected pieces of equipment meet the following:

- The total cooling load, composed of both the sensible load (temperature) and the latent load (humidity)
- The heating load
- The blower capacity (in cfm) to meet the volume of air cfm range needed.

In most areas of the lower 48 United States, there will be both a cooling load and a heating load, where the sizing of the cooling equipment will dominate the equipment selection process. A system designed to satisfy only the heating load will not have enough blower power if cooling is added after installation, because of the greater pressure loss across the cooling coil. The blower must be capable of supplying sufficient cfm airflow across the cooling coil.

Selecting equipment based on the cooling load may lead to a furnace that is oversized for the heating load. Manual S recommends that the furnace size remains within 140% of the peak heating load estimate and that cooling equipment capacity not exceed the total load by more than 15%. Manual S recommends that a heat pump in a cooling dominated climate not exceed the load by more than 15% in either the cooling or the heating mode; however, in a heating dominated climate, a heat pump may exceed the heating load by as much as 25%.

It should be recognized that in energy-efficient houses, the loads may be so small that the standard manufacturer's range of capacities exceeds these percentages. For example, if a house has a 19,000 Btu/h cooling load and manufacturer's equipment is available in only 17,500 Btu/h and 23,400 Btu/h, the larger unit is more than 15% oversized. However, the larger unit should be the equipment selected to ensure the peak load will be met. Designers may want to consider multispeed equipment in these cases, where the system runs more frequently at a lower capacity to more closely meet the load for more times of the year.

2.3. Manufacturer Data Tables

HVAC equipment manufacturers make available tabulated performance data that are used to select the equipment based on the loads and design conditions. A fictitious example of a detailed cooling capacities table is shown in Figure 6.

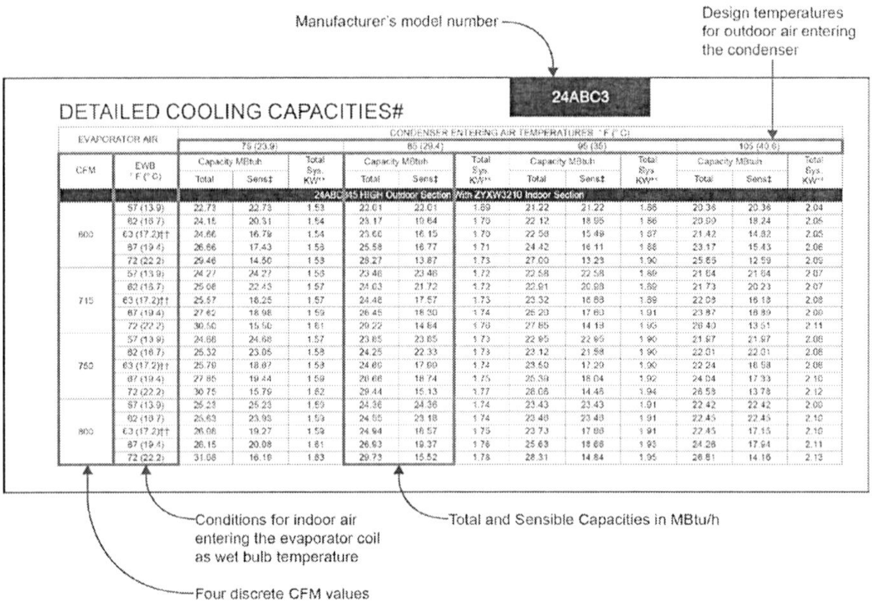

DETAILED COOLING CAPACITIES#

EVAPORATOR AIR		CONDENSER ENTERING AIR TEMPERATURES °F (°C)											
		75 (23.9)			85 (29.4)			95 (35)			105 (40.6)		
CFM	EWB °F (°C)	Capacity MBtuh Total	Capacity MBtuh Sens‡	Total Sys. KW**	Capacity MBtuh Total	Capacity MBtuh Sens‡	Total Sys. KW**	Capacity MBtuh Total	Capacity MBtuh Sens‡	Total Sys. KW**	Capacity MBtuh Total	Capacity MBtuh Sens‡	Total Sys. KW**
		24ABC345 HIGH Outdoor Section With ZYXW3210 Indoor Section											
600	57 (13.9)	22.73	22.73	1.53	22.01	22.01	1.69	21.22	21.22	1.86	20.36	20.36	2.04
	62 (16.7)	24.15	20.31	1.54	23.17	19.64	1.70	22.12	18.95	1.86	20.99	18.24	2.05
	63 (17.2)††	24.66	16.79	1.54	23.60	16.15	1.70	22.58	15.49	1.87	21.42	14.82	2.05
	67 (19.4)	26.66	17.43	1.56	25.58	16.77	1.71	24.42	16.11	1.88	23.17	15.43	2.06
	72 (22.2)	29.46	14.50	1.53	28.27	13.87	1.73	27.00	13.23	1.90	25.65	12.59	2.09
715	57 (13.9)	24.27	24.27	1.56	23.46	23.46	1.72	22.58	22.58	1.89	21.64	21.64	2.07
	62 (16.7)	25.08	22.43	1.57	24.03	21.72	1.72	22.91	20.98	1.89	21.73	20.23	2.07
	63 (17.2)††	25.57	18.25	1.57	24.48	17.57	1.73	23.32	16.88	1.89	22.08	16.18	2.08
	67 (19.4)	27.62	18.98	1.59	26.45	18.30	1.74	25.20	17.60	1.91	23.87	16.89	2.09
	72 (22.2)	30.50	15.50	1.61	29.22	14.84	1.76	27.85	14.18	1.93	26.40	13.51	2.11
750	57 (13.9)	24.68	24.68	1.57	23.85	23.85	1.73	22.95	22.95	1.90	21.97	21.97	2.08
	62 (16.7)	25.32	23.05	1.58	24.25	22.33	1.73	23.12	21.58	1.90	22.01	22.01	2.08
	63 (17.2)††	25.79	18.87	1.58	24.60	17.90	1.74	23.60	17.20	1.90	22.24	18.58	2.08
	67 (19.4)	27.85	19.44	1.59	26.66	18.74	1.75	25.39	18.04	1.92	24.04	17.33	2.10
	72 (22.2)	30.75	15.79	1.62	29.44	15.13	1.77	28.06	14.46	1.94	26.58	13.78	2.12
800	57 (13.9)	25.23	25.23	1.59	24.36	24.36	1.74	23.43	23.43	1.91	22.42	22.42	2.09
	62 (16.7)	25.63	23.93	1.59	24.55	23.18	1.74	23.48	23.48	1.91	22.45	22.45	2.10
	63 (17.2)††	26.08	19.27	1.59	24.94	18.57	1.75	23.73	17.88	1.91	22.45	17.15	2.10
	67 (19.4)	28.15	20.08	1.61	26.93	19.37	1.76	25.63	18.66	1.93	24.28	17.94	2.11
	72 (22.2)	31.08	16.18	1.63	29.73	15.52	1.78	28.31	14.84	1.95	26.81	14.16	2.13

Figure 6. Fictitious example: detailed cooling capacities table.

It is imperative that actual manufacturer's performance data are used to select equipment rather than data from the Air-Conditioning, Heating, and Refrigeration Institute (AHRI), formerly known as Air Conditioning and Refrigeration Institute (ARI) data. AHRI certification data are tested at one set of conditions to offer an "apples-to-apples" comparison of equipment performance; this is not the same set of conditions as the Manual J indoor design conditions. Also, AHRI certification data list only total capacity and not sensible or latent capacity. Therefore, AHRI does not provide enough information to accurately select a piece of equipment. Details about the limitations of AHRI data are provided in Section 8 of Manual S.

Each manufacturer may tabulate the equipment performance data differently. Taking time to understand how the equipment data are formatted is important to making the correct selection. In the example detailed cooling capacities table shown in Figure 6, the manufacturer lists the total and sensible capacities of the equipment in MBtu/h. Design temperatures for outdoor air entering the condenser are listed in a 10-degree spread, requiring interpolation to find performance at a design temperature between the listed temperatures. Interpolation of data is acceptable to determine the performance standards of specific design temperatures; extrapolation of non-linear data is not an acceptable practice. The conditions for indoor air entering the evaporator coil are listed as the entering wet-bulb (EWB) temperature.

The outside condenser and indoor coil package are selected from the manufacturer's expanded product performance data tables. The outside condenser and indoor coil package will determine the system size and operating airflows. The initial cfm airflow estimate and operating cfm will likely be different values.

2.4. Cooling Equipment Capacity Sensitivities

Capacities of cooling equipment are conditional based on the following operating conditions:

- Cubic feet per minute airflow across the coil (Figure 7)
- Outdoor air temperature (OAT) at the condenser (Figure 8)
- Indoor wet-bulb temperature at the coil (Figure 9)
- Indoor dry-bulb temperature at the coil (Figure 10).

Figures 7 through 10 show the conditional nature of direct expansion cooling equipment and that the available latent capacity is a direct relationship to the total and sensible capacities.

It is helpful to consider these relationships while searching the manufacturer's detailed cooling capacity data. This conditional nature among total, sensible, and latent capacities causes an imbalance between latent load and latent capacity, where latent capacity slightly exceeds the latent load. This can be somewhat self-correcting. Manual S, Section 3-6, explains that with a latent capacity that slightly exceeds the latent load, about half the excess latent capacity will be converted to sensible capacity, aiding in equipment selection.

Figure 7 shows that increasing the airflow across the coil (cfm) will increase the total and sensible capacities of the equipment. However, because the sensible capacity increases at a higher rate than the total capacity, the latent capacity available is reduced as the cfm increases. This relationship between cfm and latent capacity of the equipment is an important consideration when selecting equipment to meet the latent load.

Figure 8 shows that as the OAT at the condenser increases, the total sensible and latent capacities decrease. Because the total capacity decreases at a faster rate than the sensible capacity, the available latent capacity of the equipment is lower. Because of this effect of OAT on equipment capacity, it is critical that the same temperature used for load calculations is used for equipment selection.

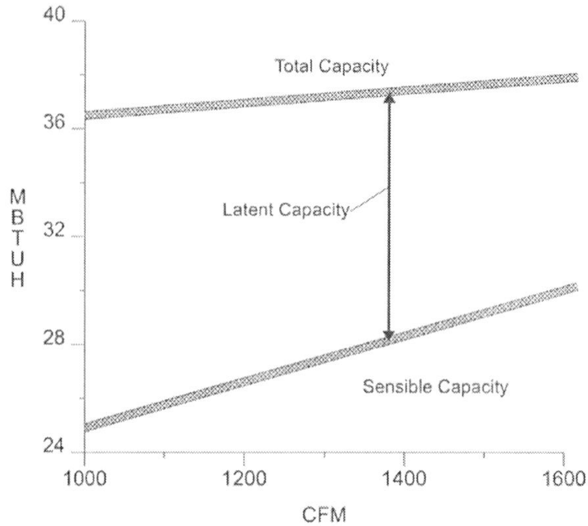

Figure 7. Airflow sensitivity graph (from Manual S, Figure 3-1). (*Reprinted with permission from Air Conditioning Contractors of America*).

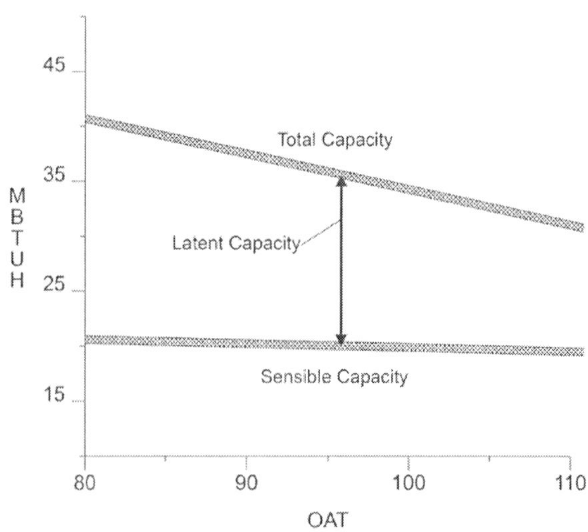

Figure 8. OAT sensitivity graph (from Manual S, Figure 3-1). (*Reprinted with permission from Air Conditioning Contractors of America*).

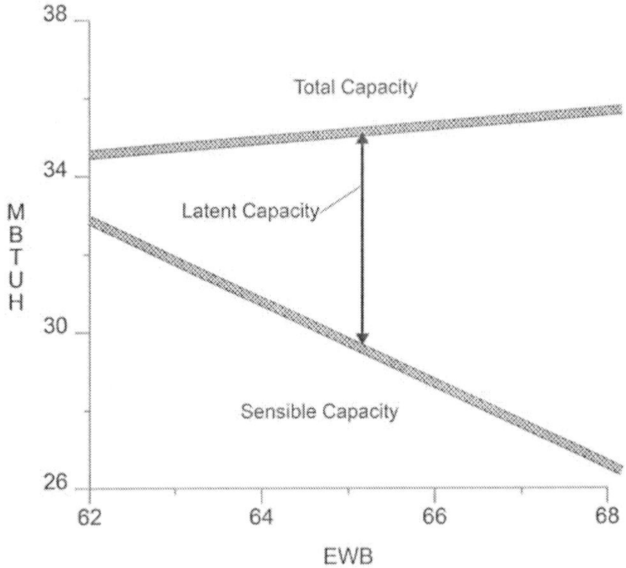

Figure 9. Indoor EWB sensitivity graph (from Manual S, Figure 3-1). (*Reprinted with permission from Air Conditioning Contractors of America*).

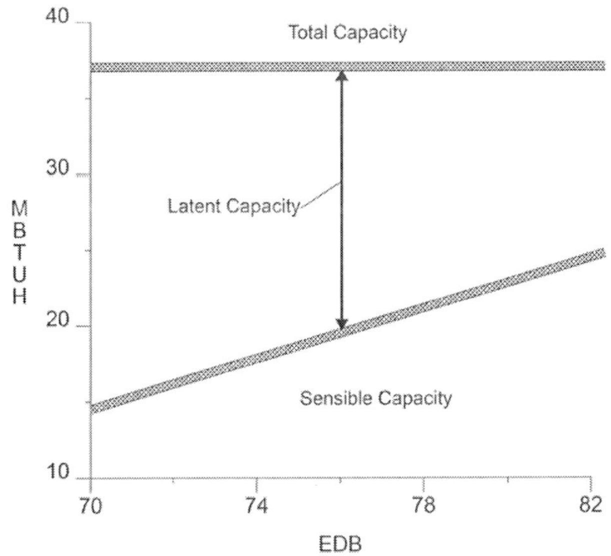

Figure 10. Indoor EDB temperature sensitivity graph (from Manual S, Figure 3-1). (*Reprinted with permission from Air Conditioning Contractors of America*).

As shown in Figure 9, higher EWB temperatures at the indoor coil will result in a higher total capacity but a lower sensible capacity. This will result in a greater available latent capacity as the difference between the total and sensible capacities increases.

Figure 10 shows that a higher entering dry-bulb (EDB) temperature at the indoor coil will have relatively little effect on the total capacity but will increase the sensible capacity of the equipment, resulting in a lower available latent capacity.

2.5. Estimated Target Cubic Feet per Minute

To search the manufacturer's expanded performance data, an initial estimate of the required cfm airflow is needed. The estimated cfm airflow is used only to narrow down the equipment choices from the manufacturer's data and will likely be higher than the actual cfm used for the design.

The detailed procedure to estimate a target cfm is described in Manual S. Section 1.3 of Manual S instructs the designer to use the sensible heat ratio (SHR) value to determine the desired room-air to supply-air temperature difference (TD) in degrees Fahrenheit. The SHR is the ratio of the sensible load determined from the Manual J procedures to the Manual J total load shown in Equation 1.

Sensible Heat Ratio (SHR) – The sensible load divided by the total load.

Temperature Difference (TD) – The difference between the room air temperature and the supply air temperature.

$$\text{SHR} = \frac{\text{Sensible Load}}{\text{Total Load}} \qquad (1)$$

Figure 11, taken from Table 1-4 in Manual S, shows the relationship of an SHR range to TD value that is compatible with the sensible and latent loads. The column "LAT" is the temperature of air leaving the coil, and the column "Room db" is the indoor conditions dry-bulb temperature used in the Manual J calculations. The difference between the two is the TD value.

Sensible Heat Ratio Versus TD Value			
SHR	LAT	Room db	TD
Below 0.80	54	75	21
0.80 to 0.85	56	75	19
Above 0.85	58	75	17

Figure 11. TD value table (from Manual S, Table 1-4). (*Reprinted with permission from Air Conditioning Contractors of America*).

Determining the SHR early is helpful because the SHR can also be used as a double check on the accuracy of the Manual J loads. A relatively high SHR (e.g., above 0.85) indicates a relatively small latent load compared to the sensible load. Conversely, a relatively low SHR (e.g., below 0.75) indicates the latent load is relatively large compared to the sensible load.

Once the TD value is known, the estimated cfm target can be calculated using Equation 2.

$$\text{cfm Target} = \frac{\text{Manual J Sensible Load}}{1.1 \times \text{TD}} \tag{2}$$

The number 1.1 in the denominator of Equation 2 represents an air properties constant associated with elevations lower than 1,500 feet and is appropriate for both the Chicago House and the Orlando House examples. The value of the air properties constant will decrease with higher elevations. Correction factor information can be found in Appendix 6 of Manual S.

2.6. Efficiency Considerations

The selection of the efficiency rating (AFUE or SEER) of the equipment will affect the long-term energy consumption and system operating costs. Although the decision-making process for selecting an efficiency level is a cost/benefit analysis of installed first cost versus the operational savings of the more efficient equipment, it is anticipated that future legislation will tighten the efficiency requirements of HVAC equipment, making more efficient equipment mandatory. To illustrate the impact of equipment efficiency on operating energy use and costs, two performance levels were compared using Building Energy Optimization software, version 1.1 (BEopt 1.1).

For the Chicago House example, an 80% AFUE furnace with a 13 SEER air conditioner was compared to a 92.5% AFUE furnace with a 16 SEER air conditioner. The results show a total annual source energy use reduction of 37.7 MBtu/yr or 11%. The bulk of the reduction, 31.6 MBtu/yr, was in the heating mode with 1.7 MBtu/yr in cooling and the remaining 4.4 MBtu/yr reduction coming from fan energy savings. Figure 12 shows that, on an annualized utility bill basis (dollars/year), the more efficient equipment in the Chicago House resulted in an estimated $273 reduction.

For the Orlando House, an 8.0 HSPF/13 SEER heat pump was compared with a 9.2 HSPF/16 SEER heat pump. The results show a total annual source energy use reduction of 9.3 MBtu/yr or 13%. The bulk of the reduction, 7.5 MBtu/yr, was in the cooling mode with 0.7 MBtu/yr in heating and the remaining 1.1 MBtu/yr reduction coming from fan energy savings. Figure 13 shows that, on an annualized utility bill basis (dollars/year), the more efficient equipment in the Orlando House resulted in an estimated $65 reduction.

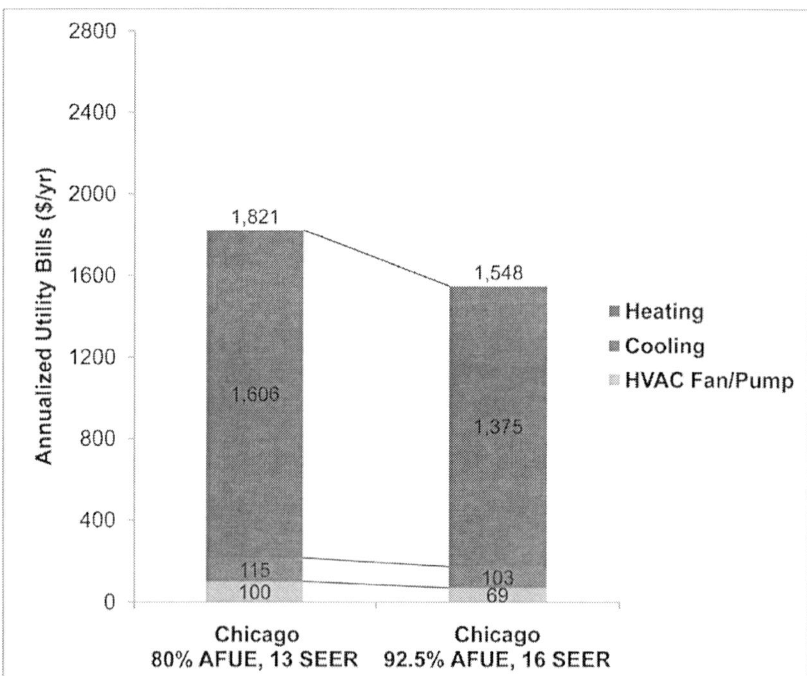

Figure 12. Chicago House BEopt 1.1 output.

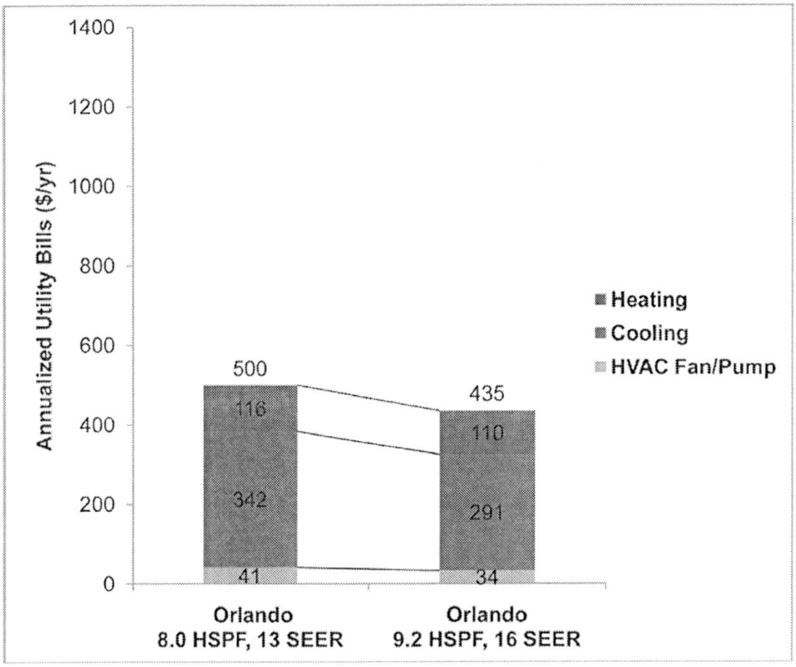

Figure 13. Orlando House BEopt 1.1 output.

3.1. Cooling Equipment Selection

Split system cooling equipment consisting of an outdoor condensing unit and an indoor evaporator coil needs the two pieces to be matched for efficiency and performance. The example here shows a manufacturer matched equipment package. Although it is possible to select the two pieces separately, it is typically not necessary. The manufacturer matched package greatly simplifies the cooling equipment selection.

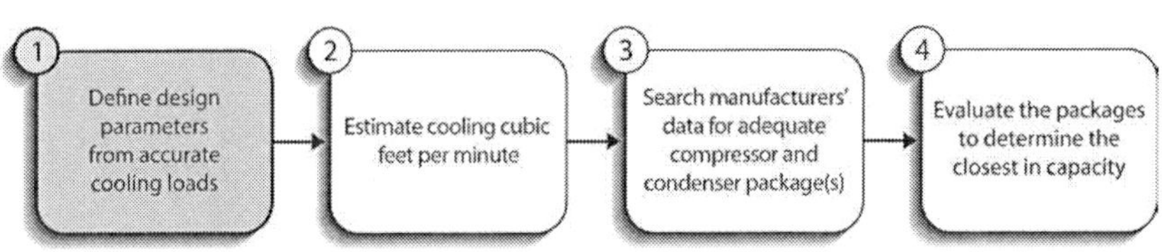

Figure 14. Cooling equipment selection step 1.

Manual J	Heating Load	41,660 Btu/h
	Sensible Cooling Load	17,370 Btu/h
	Latent Cooling Load	3,240 Btu/h
Outdoor Conditions	Cooling Dry-bulb Temperature	89 °F
	Heating Dry-bulb Temperature	2 °F
Indoor Conditions	Cooling Dry-bulb Temperature	75 °F
	Cooling Wet-bulb Temperature	63 °F
	Heating Dry-bulb Temperature	70 °F
Airflow Estimates	Cubic Feet per Minute	831 cfm

Figure 15. Chicago House design information.

The first step highlighted in Figure 14 instructs the designer to define the cooling design parameters from the cooling loads of the Manual J procedure. The cooling design information developed by Burdick (2011) from the Manual J procedures is shown in Figure 15.

For the example houses in this Strategy Guide, a 92.5% AFUE furnace with a 16 SEER air conditioner was the system chosen for the Chicago House and a 9.2 HSPF/16 SEER heat pump for the Orlando House.

3. Chicago House Example

The following is an example selection process of a split system for the Chicago House with an upflow gas furnace located in the basement and an air conditioner. The Chicago House has both a cooling load and a heating load; therefore, the cooling equipment will be selected first.

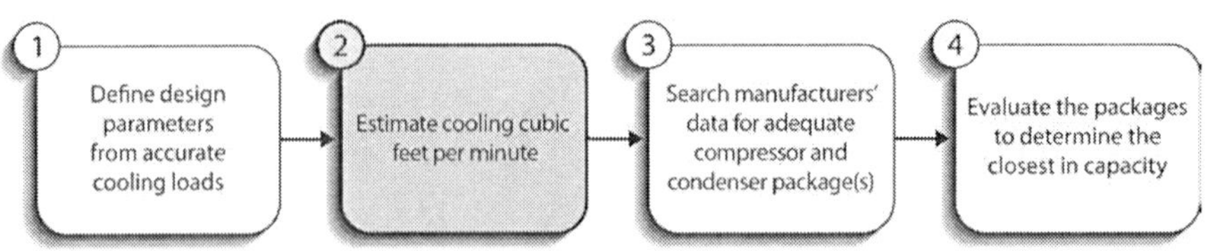

Figure 16. Cooling equipment selection step 2.

Step 2, highlighted in Figure 16, instructs the designer to estimate the cooling cfm. Utilizing the methods outlined in Section 1.3 of Manual S, the initial cfm airflow is estimated.

Equation 3 applies the cooling design parameters from Figure 15 to Equation 1, resulting in an SHR of 0.84.

$$\text{SHR} = \frac{\text{Sensible Load}}{\text{Total Load}} = \frac{17.37\frac{\text{MBtu}}{\text{h}}}{20.61\frac{\text{MBtu}}{\text{h}}} = 0.84 \tag{3}$$

From the TD value table (Figure 11), the TD value for SHR = 0.84 is 19. These values can now be applied to the sensible heat equation (Equation 2) as shown in Equation 4.

$$\text{cfm Target} = \frac{\text{Manual J Sensible Load}}{1.1 \times \text{TD}} = \frac{17{,}370}{1.1 \times 19} = 831 \tag{4}$$

Step 3, highlighted in Figure 17, is to search the manufacturer's data for an adequate cooling equipment package.

To search the fictitious example of a cooling equipment package, detailed cooling capacities, as shown in Figure 6, a nominal size must be determined. Considering that 12.00 MBtu/h is equivalent to 1 nominal ton of cooling, the 20,610 Btu/h total cooling load for the Chicago House leads to a 2-ton nominal size cooling system as the target size to begin the equipment search.

The EWB temperature for the indoor coil can be determined on the sea-level psychometric chart from the Manual J conditions of 75°F indoor design dry-bulb temperature and 50% relative humidity, as shown in Figure 18.

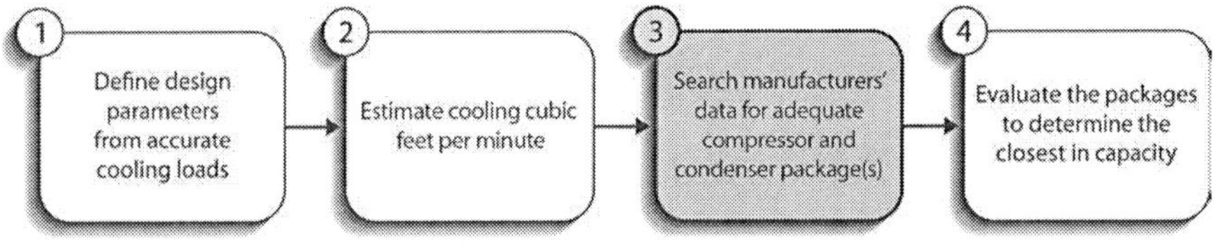

Figure 17. Cooling equipment selection step 3.

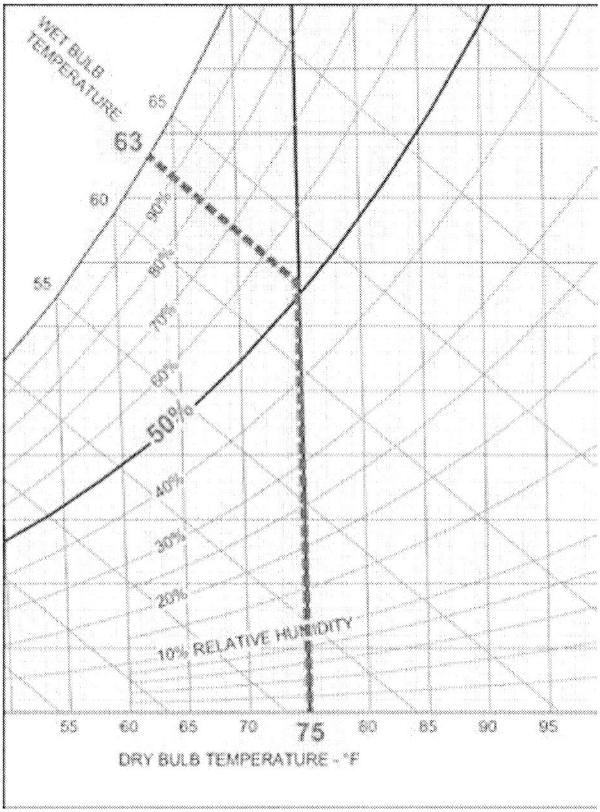

Figure 18. Wet-bulb temperature on sea-level psychometric chart.

The detailed cooling capacities tables give the data required for step 4 of Figure 19.

The outside design temperature for the Chicago House used in the Manual J load calculation was 89°F. The 89°F outdoor design temperature will require interpolation of the manufacturer's published data between the 85°F and 95°F condenser entering air temperature points.

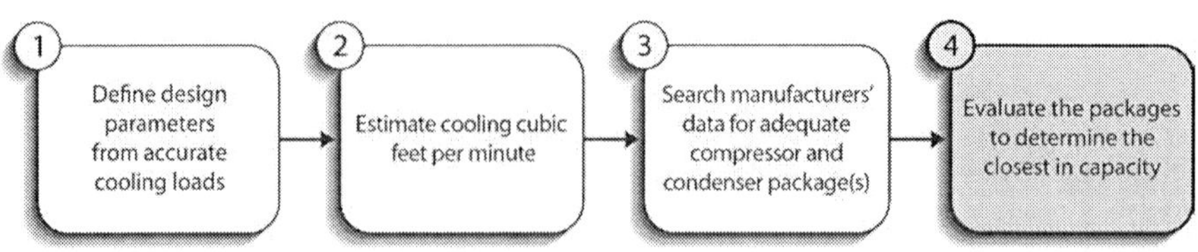

Figure 19. Cooling equipment selection step 4.

DETAILED COOLING CAPACITIES# — 24ABC3

EVAPORATOR AIR		CONDENSER ENTERING AIR TEMPERATURES °F (°C)											
		75 (23.9)			85 (29.4)			95 (35)			105 (40.6)		
CFM	EWB °F (°C)	Capacity MBtuh Total	Capacity MBtuh Sens‡	Total Sys. KW**	Capacity MBtuh Total	Capacity MBtuh Sens‡	Total Sys. KW**	Capacity MBtuh Total	Capacity MBtuh Sens‡	Total Sys. KW**	Capacity MBtuh Total	Capacity MBtuh Sens‡	Total Sys. KW**
24ABC345 HIGH Outdoor Section With ZYXW3210 Indoor Section													
[illegible]	[illegible]	[illegible]	[illegible]	[illegible]	[illegible]	[illegible]	[illegible]	[illegible]	[illegible]	[illegible]	[illegible]	[illegible]	[illegible]
	[illegible]	[illegible]	[illegible]	[illegible]	[illegible]	[illegible]	[illegible]	[illegible]	[illegible]	[illegible]	[illegible]	[illegible]	[illegible]
	[illegible]	[illegible]	[illegible]	[illegible]	[illegible]	[illegible]	[illegible]	[illegible]	[illegible]	[illegible]	[illegible]	[illegible]	[illegible]
	[illegible]	[illegible]	[illegible]	[illegible]	[illegible]	[illegible]	[illegible]	[illegible]	[illegible]	[illegible]	[illegible]	[illegible]	[illegible]
	[illegible]	[illegible]	[illegible]	[illegible]	[illegible]	[illegible]	[illegible]	[illegible]	[illegible]	[illegible]	[illegible]	[illegible]	[illegible]
[illegible]	[illegible]	[illegible]	[illegible]	[illegible]	[illegible]	[illegible]	[illegible]	[illegible]	[illegible]	[illegible]	[illegible]	[illegible]	[illegible]
	[illegible]	[illegible]	[illegible]	[illegible]	[illegible]	[illegible]	[illegible]	[illegible]	[illegible]	[illegible]	[illegible]	[illegible]	[illegible]
	63 (17.2)††	[illegible]	[illegible]	[illegible]	24.46	17.57	[illegible]	23.32	16.88	[illegible]	[illegible]	[illegible]	[illegible]
	[illegible]	[illegible]	[illegible]	[illegible]	[illegible]	[illegible]	[illegible]	[illegible]	[illegible]	[illegible]	[illegible]	[illegible]	[illegible]
	[illegible]	[illegible]	[illegible]	[illegible]	[illegible]	[illegible]	[illegible]	[illegible]	[illegible]	[illegible]	[illegible]	[illegible]	[illegible]
[illegible]	[illegible]	[illegible]	[illegible]	[illegible]	[illegible]	[illegible]	[illegible]	[illegible]	[illegible]	[illegible]	[illegible]	[illegible]	[illegible]
	[illegible]	[illegible]	[illegible]	[illegible]	[illegible]	[illegible]	[illegible]	[illegible]	[illegible]	[illegible]	[illegible]	[illegible]	[illegible]
	63 (17.2)††	[illegible]	[illegible]	[illegible]	24.69	17.99	[illegible]	23.59	17.29	[illegible]	[illegible]	[illegible]	[illegible]
	[illegible]	[illegible]	[illegible]	[illegible]	[illegible]	[illegible]	[illegible]	[illegible]	[illegible]	[illegible]	[illegible]	[illegible]	[illegible]
	[illegible]	[illegible]	[illegible]	[illegible]	[illegible]	[illegible]	[illegible]	[illegible]	[illegible]	[illegible]	[illegible]	[illegible]	[illegible]
800	[illegible]	[illegible]	[illegible]	[illegible]	[illegible]	[illegible]	[illegible]	[illegible]	[illegible]	[illegible]	[illegible]	[illegible]	[illegible]
	[illegible]	[illegible]	[illegible]	[illegible]	[illegible]	[illegible]	[illegible]	[illegible]	[illegible]	[illegible]	[illegible]	[illegible]	[illegible]
	63 (17.2)††	[illegible]	[illegible]	[illegible]	24.94	18.57	[illegible]	23.73	17.86	[illegible]	[illegible]	[illegible]	[illegible]
	[illegible]	[illegible]	[illegible]	[illegible]	[illegible]	[illegible]	[illegible]	[illegible]	[illegible]	[illegible]	[illegible]	[illegible]	[illegible]
	[illegible]	[illegible]	[illegible]	[illegible]	[illegible]	[illegible]	[illegible]	[illegible]	[illegible]	[illegible]	[illegible]	[illegible]	[illegible]

Figure 20. Fictitious example: detailed cooling capacities table, 800 cfm.

The fictitious example cooling equipment package detailed cooling capacities table is searched at the closest published cfm data to the initial cfm airflow estimate of 831 cfm. Figure 20 highlights the 800 cfm and 63°F EWT temperature row, depicting a total capacity of 24.94 MBtu/h at 85°F and 23.73 MBtu/h at 95°F.

It is clear without going through the interpolation exercise that this cooling equipment total capacity at 800 cfm exceeds the sizing limits for the total Manual J cooling load of 15% larger than the peak total load (see Equation 5).

$$20.61\,\text{MBtu/h} \times 1.15 - 23.70\,\text{MBtu/h} \tag{5}$$

This is not an unexpected result because the initial 831 cfm airflow estimate is in alignment with the 400 cfm per 12.00 MBtu/h conventional methodologies of lower performing equipment and less efficient houses. It is not uncommon to find the target cfm range for a higher efficiency house with higher efficiency equipment to fall closer to 350 cfm per 12.00 MBtu/h.

Evaluating the same piece of equipment at lower cfm is the appropriate next step. The 715 cfm data row in Figure 21 shows the total capacity to be closer to the total load requirement. However, the sensible capacity values of 17.57 MBtu/h at 85°F and 16.88 MBtu/h at 95°F show, without interpolating between the values, that the package will likely not meet the sensible portion of the load (17.37 MBtu/h) at 89°F.

DETAILED COOLING CAPACITIES# — **24ABC3**

EVAPORATOR AIR		CONDENSER ENTERING AIR TEMPERATURES °F (°C)											
		75 (23.9)			85 (29.4)			95 (35)			105 (40.6)		
CFM	EWB °F (°C)	Capacity MBtuh Total	Capacity MBtuh Sens‡	Total Sys. KW**	Capacity MBtuh Total	Capacity MBtuh Sens‡	Total Sys. KW**	Capacity MBtuh Total	Capacity MBtuh Sens‡	Total Sys. KW**	Capacity MBtuh Total	Capacity MBtuh Sens‡	Total Sys. KW**
24ABC345 HIGH Outdoor Section With ZYXW3210 Indoor Section													
600	[illegible]	[illegible]	[illegible]	[illegible]	[illegible]	[illegible]	[illegible]	[illegible]	[illegible]	[illegible]	[illegible]	[illegible]	[illegible]
	[illegible]	[illegible]	[illegible]	[illegible]	[illegible]	[illegible]	[illegible]	[illegible]	[illegible]	[illegible]	[illegible]	[illegible]	[illegible]
	[illegible]	[illegible]	[illegible]	[illegible]	[illegible]	[illegible]	[illegible]	[illegible]	[illegible]	[illegible]	[illegible]	[illegible]	[illegible]
	[illegible]	[illegible]	[illegible]	[illegible]	[illegible]	[illegible]	[illegible]	[illegible]	[illegible]	[illegible]	[illegible]	[illegible]	[illegible]
	[illegible]	[illegible]	[illegible]	[illegible]	[illegible]	[illegible]	[illegible]	[illegible]	[illegible]	[illegible]	[illegible]	[illegible]	[illegible]
715	[illegible]	[illegible]	[illegible]	[illegible]	[illegible]	[illegible]	[illegible]	[illegible]	[illegible]	[illegible]	[illegible]	[illegible]	[illegible]
	[illegible]	[illegible]	[illegible]	[illegible]	[illegible]	[illegible]	[illegible]	[illegible]	[illegible]	[illegible]	[illegible]	[illegible]	[illegible]
	63 (17.2)††	[illegible]	[illegible]	[illegible]	24.48	17.57	[illegible]	23.32	16.88	[illegible]	[illegible]	[illegible]	[illegible]
	[illegible]	[illegible]	[illegible]	[illegible]	[illegible]	[illegible]	[illegible]	[illegible]	[illegible]	[illegible]	[illegible]	[illegible]	[illegible]
	[illegible]	[illegible]	[illegible]	[illegible]	[illegible]	[illegible]	[illegible]	[illegible]	[illegible]	[illegible]	[illegible]	[illegible]	[illegible]
750	[illegible]	[illegible]	[illegible]	[illegible]	[illegible]	[illegible]	[illegible]	[illegible]	[illegible]	[illegible]	[illegible]	[illegible]	[illegible]
	[illegible]	[illegible]	[illegible]	[illegible]	[illegible]	[illegible]	[illegible]	[illegible]	[illegible]	[illegible]	[illegible]	[illegible]	[illegible]
	63 (17.2)††	[illegible]	[illegible]	[illegible]	24.69	17.99	[illegible]	23.50	17.29	[illegible]	[illegible]	[illegible]	[illegible]
	[illegible]	[illegible]	[illegible]	[illegible]	[illegible]	[illegible]	[illegible]	[illegible]	[illegible]	[illegible]	[illegible]	[illegible]	[illegible]
	[illegible]	[illegible]	[illegible]	[illegible]	[illegible]	[illegible]	[illegible]	[illegible]	[illegible]	[illegible]	[illegible]	[illegible]	[illegible]
[illegible]	[illegible]	[illegible]	[illegible]	[illegible]	[illegible]	[illegible]	[illegible]	[illegible]	[illegible]	[illegible]	[illegible]	[illegible]	[illegible]
	[illegible]	[illegible]	[illegible]	[illegible]	[illegible]	[illegible]	[illegible]	[illegible]	[illegible]	[illegible]	[illegible]	[illegible]	[illegible]
	63 (17.2)††	[illegible]	[illegible]	[illegible]	24.84	18.57	[illegible]	23.73	17.86	[illegible]	[illegible]	[illegible]	[illegible]
	[illegible]	[illegible]	[illegible]	[illegible]	[illegible]	[illegible]	[illegible]	[illegible]	[illegible]	[illegible]	[illegible]	[illegible]	[illegible]
	[illegible]	[illegible]	[illegible]	[illegible]	[illegible]	[illegible]	[illegible]	[illegible]	[illegible]	[illegible]	[illegible]	[illegible]	[illegible]

Figure 21. Fictitious example: detailed cooling capacities table, 715 cfm and 750 cfm.

City: Chicago, IL	Cooling Outdoor Design Temperature = 89 °F										
	Total Cooling Design Load = 20.6 MBtu/h										
	Sensible Cooling Design Load = 17.4 MBtu/h										
	Condenser Entering Air Temperatures °F										
	85	86	87	88	89	90	91	92	93	94	95
Total Cooling (CFM)	Total Capacity (MBtu/h)										
600	23.66	23.55	23.44	23.34	23.23	23.12	23.01	22.90	22.80	22.69	22.58
715	24.45	24.33	24.22	24.10	23.98	23.87	23.75	23.63	23.52	23.40	23.29
750	24.69	24.57	24.45	24.33	24.21	24.10	23.98	23.86	23.74	23.62	23.50
	Sensible Capacity (MBtu/h)										
750	17.99	17.92	17.85	17.78	17.71	17.64	17.57	17.50	17.43	17.36	17.29

Figure 22. Interpolation table for capacities at 89°F outdoor design temperature.

The published 750 cfm data in Figure 21 for both total and sensible capacities appear to be within the range of the load requirements. Figure 22 shows the linear interpolation of the values to determine the capacity at 89°F. This analysis shows that the capacity of this package at 89°F will meet the total, sensible, and latent loads at 750 cfm. Utilizing a spreadsheet to handle the multiple interpolation calculations makes the procedure more manageable (see Figure 23).

The fictitious example cooling equipment package will exceed our cooling load requirements by 17% at 750 cfm. As previously stated, in energy efficient houses the loads may be so small that the standard manufacturer's range of capacities could exceed the 15% sizing limits. Equipment must be selected to meet the peak load without grossly oversizing the equipment, and the selection should be based on the selection procedures backed up by the data.

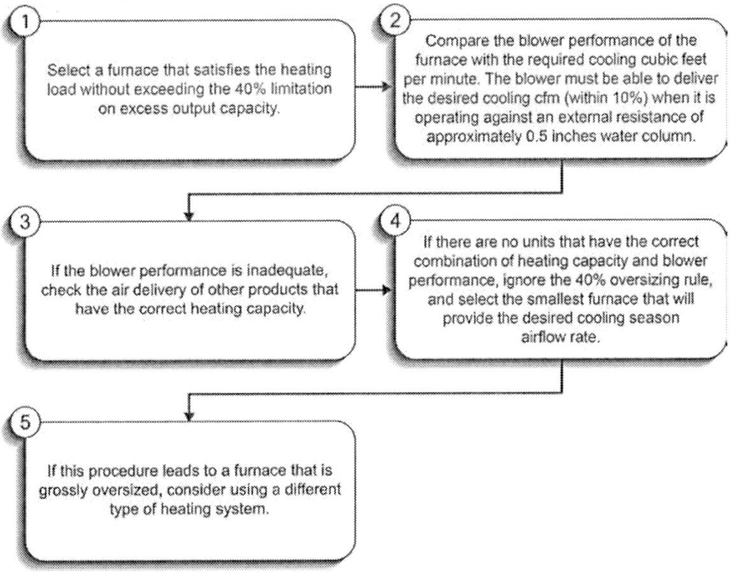

Figure 23. Furnace selection steps.

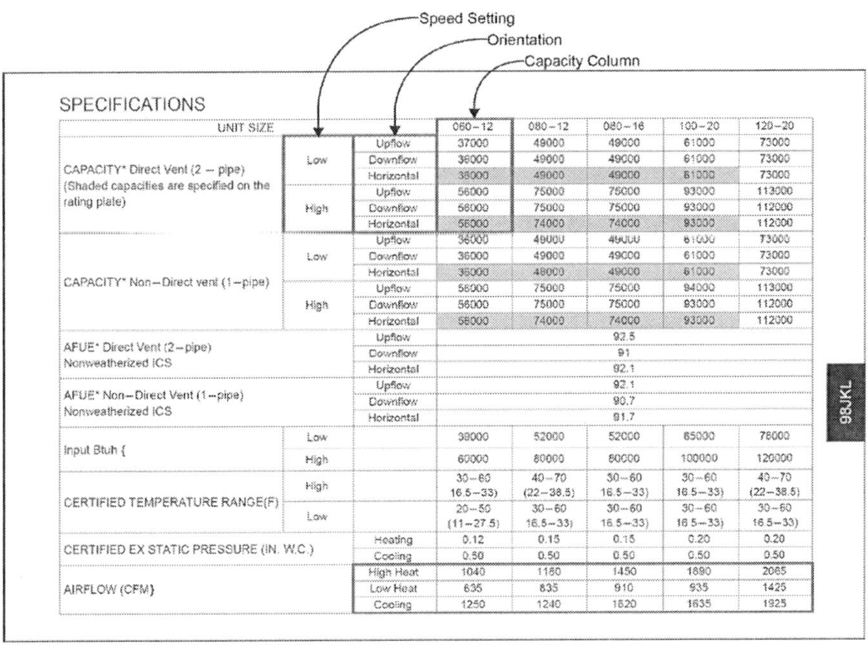

SPECIFICATIONS

UNIT SIZE			060–12	080–12	080–16	100–20	120–20
CAPACITY* Direct Vent (2 – pipe) (Shaded capacities are specified on the rating plate)	Low	Upflow	37000	49000	49000	61000	73000
		Downflow	36000	49000	49000	61000	73000
		Horizontal	36000	49000	49000	61000	73000
	High	Upflow	56000	75000	75000	93000	113000
		Downflow	56000	75000	75000	93000	112000
		Horizontal	56000	74000	74000	93000	112000
CAPACITY* Non–Direct vent (1–pipe)	Low	Upflow	36000	49000	49000	61000	73000
		Downflow	36000	49000	49000	61000	73000
		Horizontal	36000	48000	49000	61000	73000
	High	Upflow	56000	75000	75000	94000	113000
		Downflow	56000	75000	75000	93000	112000
		Horizontal	56000	74000	74000	93000	112000
AFUE* Direct Vent (2–pipe) Nonweatherized ICS		Upflow	92.5				
		Downflow	91				
		Horizontal	92.1				
AFUE* Non–Direct Vent (1–pipe) Nonweatherized ICS		Upflow	92.1				
		Downflow	90.7				
		Horizontal	91.7				
Input Btuh {	Low		39000	52000	52000	65000	78000
	High		60000	80000	80000	100000	120000
CERTIFIED TEMPERATURE RANGE(F)	High		30–60 16.5–33)	40–70 (22–38.5)	30–60 16.5–33)	30–60 16.5–33)	40–70 (22–38.5)
	Low		20–50 (11–27.5)	30–60 16.5–33)	30–60 16.5–33)	30–60 16.5–33)	30–60 16.5–33)
CERTIFIED EX STATIC PRESSURE (IN. W.C.)		Heating	0.12	0.15	0.15	0.20	0.20
		Cooling	0.50	0.50	0.50	0.50	0.50
AIRFLOW (CFM)		High Heat	1040	1180	1450	1890	2065
		Low Heat	635	835	910	935	1425
		Cooling	1250	1240	1620	1635	1925

Figure 24. Furnace specification chart.

The target 750 cfm value will be carried through the remainder of the system design procedures; however, this cfm value may be revisited and changed when the duct design is done. A furnace for the Chicago House can now be selected based on the cooling equipment selection and the Manual J heating loads.

3.2. Furnace Selection

With the cooling equipment package selected, a furnace must be selected that meets the heating load. The furnace capacity must meet the Manual J heating load while not exceeding the heating load by more than 40%. From Manual S, Section 2-5, the steps to selecting a furnace are shown in Figure 23.

Figure 24 shows that furnace capacities span a much larger range than cooling equipment capacities. A particular manufacturer may size a furnace line from 60,000 Btu/h to 120,000 Btu/h with 80,000 Btu/h as the intermediate size. This can be alleviated somewhat with multiple-stage equipment. A two-stage furnace in high stage may have a rating of 60,000 Btu/h but can be 39,000 Btu/h in low stage. Furnaces are rated as the input Btu/h, but the capacity is derated by the efficiency level. For example, a 60,000 Btu/h furnace with a 92.5% efficiency rating will yield around 55,500 Btu/h output capacity.

Each manufacturer may tabulate the furnace performance data differently. The furnace must also dimensionally match the size of the indoor coil. The example equipment manufacturer furnace is matched to the indoor cooling coil size and type, but not all manufacturers treat coil matching the same way. Some may require that the designers verify the furnace and coil are dimensionally compatible. Taking time to understand how the furnace data are formatted is important to making the correct selection.

Figure 25 is an example of an air delivery performance table. The blower performance of the furnace and the required cooling cfm must be compared. While comparing the blower performance of the furnace and the required cooling cfm, the blower must be able to meet the cooling cfm within 10%. The assumption at the start of the equipment selection process is 0.50 inch water column external static pressure for the system. The equipment should have the capability of field setting the blower speed to match the cooling cfm per the manufacturer's instructions.

To select a furnace for the Chicago House, the heating load estimate from the Manual J procedure of 41,660 Btu/h must be satisfied. The two parameters that must be considered for selection are the heating capacity and the cfm airflow. Equation 6 sets the maximum target capacity of a furnace for the Chicago House.

$$41{,}660 \text{ Btu/h} \times 1.4 = 58{,}324 \text{ Btu/h} \quad (6)$$

The furnace with specifications as shown in Figure 26 will meet the required load without exceeding the maximum target capacity in the upflow configuration at the high setting. The cooling airflow of 1250 cfm almost doubles the 750 cfm design airflow determined in the cooling equipment selection; however, Figure 27 shows that the equipment is field adjustable to within the airflow margin constraint of 0.5 inch water column external static pressure.

AIR DELIVERY - CFM (WITH FILTER)

UNIT SIZE	RETURN-AIR SUPPLY	SPEED	EXTERNAL STATIC PRESSURE (IN. W.C.)									
			0.1	0.2	0.3	0.4	0.5	0.6	0.7	0.8	0.9	1
060-12	SIDE/BOTTOM	5(Gry)	1430	1390	1345	1300	1250	1205	1150	1080	995	845
		3(Blu)	1240	1200	1145	1100	1040	975	915	860	790	730
		4(Yel)	1090	1030	980	935	850	800	730	665	590	525
		1(Red)	900	835	780	705	635	565	490	410	335	200
		2(Org)	805	620	440	380	300	—	—	—	—	—

98JKL

Equipment is field adjustable to desired CFM

Design assumption is 0.50 iwc external static pressure

Figure 25. Air delivery performance table.

SPECIFICATIONS

UNIT SIZE			060-12	080-12	080-16	100-20	120-20
CAPACITY* Direct Vent (2 - pipe) (Shaded capacities are specified on the rating plate)	Low	Upflow	37000	40000	40000	61000	73000
		Downflow	36000	49000	49000	61000	73000
		Horizontal	36000	49000	49000	61000	73000
	High	Upflow	56000	75000	75000	93000	113000
		Downflow	56000	75000	75000	93000	112000
		Horizontal	56000	74000	74000	93000	112000
CAPACITY* Non-Direct vent (1-pipe)	Low	Upflow	36000	49000	49000	61000	73000
		Downflow	36000	49000	49000	61000	73000
		Horizontal	36000	48000	49000	61000	73000
	High	Upflow	56000	75000	75000	94000	113000
		Downflow	56000	75000	75000	93000	112000
		Horizontal	56000	74000	74000	93000	112000
AFUE* Direct Vent (2-pipe) Nonweatherized ICS		Upflow	92.5				
		Downflow	91				
		Horizontal	92.1				
AFUE* Non-Direct Vent (1-pipe) Nonweatherized ICS		Upflow	92.1				
		Downflow	90.7				
		Horizontal	91.7				
Input Btuh (	Low		39000	52000	52000	65000	78000
	High		60000	80000	80000	100000	120000
CERTIFIED TEMPERATURE RANGE(F)	High		30-60 16.5-33)	40-70 (22-38.5)	30-60 16.5-33)	30-60 16.5-33)	40-70 (22-38.5)
	Low		20-50 (11-27.5)	30-60 16.5-33)	30-60 16.5-33)	30-60 16.5-33)	30-60 16.5-33)
CERTIFIED EX STATIC PRESSURE (IN. W.C.)		Heating	0.12	0.15	0.15	0.20	0.20
		Cooling	0.50	0.50	0.50	0.50	0.50
AIRFLOW (CFM)		High Heat	1040	1180	1450	1890	2065
		Low Heat	635	835	910	935	1425
		Cooling	1250	1240	1620	1635	1925

98JKL

Figure 26. Fictitious furnace performance example.

AIR DELIVERY - CFM (WITH FILTER)

UNIT SIZE	RETURN-AIR SUPPLY	SPEED	EXTERNAL STATIC PRESSURE (IN. W.C.)									
			0.1	0.2	0.3	0.4	0.5	0.6	0.7	0.8	0.9	1
060-12	SIDE/BOTTOM	5(Gry)	1430	1390	1345	1300	1250	1205	1150	1080	995	845
		3(Blu)	1240	1200	1145	1100	1040	975	915	860	790	730
		4(Yel)	1090	1030	980	935	850	800	730	665	590	525
		1(Red)	900	835	780	705	635	565	490	410	335	200
		2(Org)	805	620	440	380	300	---	---	---	---	---

98JKL

Figure 27. Air delivery performance table for the Chicago House.

It is possible that conditions will arise from the duct design that may change the external static pressure and may require a different piece of equipment be chosen. The iterative nature of the equipment selection process is in balancing the capacity of the equipment to meet the load with the blower performance.

4. ORLANDO HOUSE EXAMPLE

The Orlando House example equipment choice is a 9.2 HSPF/16 SEER air source heat pump. The selection of an air source heat pump is driven by climatic conditions, the size of the cooling load compared to the heating load, and the economic factors of gas prices compared to the cost of electricity. Air source heat pump systems have not traditionally been a common choice in cold climates with a large heating load; however, an air source heat pump is a likely choice for climates where the heating load is much smaller than the cooling load. Dual fuel options can make an air source heat pump a viable choice for a cold climate by providing an economical solution of a gas-fired burner to meet the larger heating load in cold climates. This evaluation will require an in-depth cost comparison. Details of the economic balance-point evaluation for a dual fuel system are covered in Manual S, Section 6.

Initially, heat pump equipment selection is based on the cooling load. The selection of heat pump equipment to meet the cooling capacity follows the selection of cooling equipment in the air conditioner selection in Section 3 of Manual S and is sensitive to the same four operating parameters as the split system cooling equipment. In warm climates, the sizing limitations of an air source heat pump are 15% greater than the Manual J loads for heating and total cooling loads, whereas in cold climates, it is permissible to exceed the Manual J heating load by 25%.

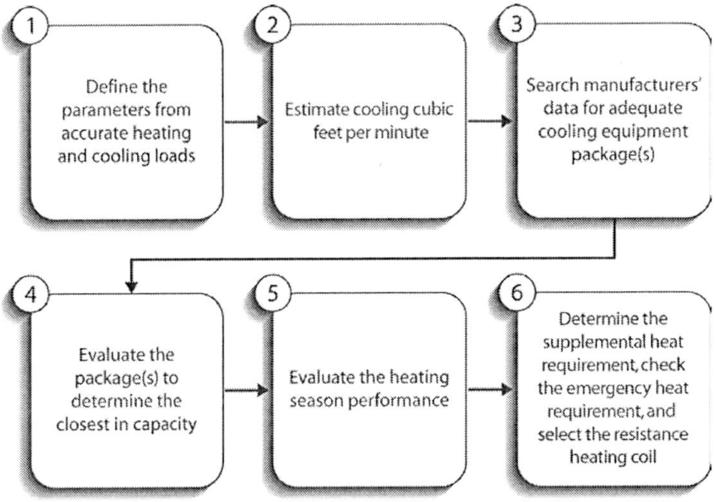

Figure 28. Air source heat pump selection steps.

Because no furnace is used in this particular heat pump system, an indoor fan coil air handler with electric resistance heat coils will be selected to match the performance requirements and design parameters for airflow and heating load. The additional information required for heat pump selection is the amount of supplemental heat and the emergency heat capacity.

Based on Section 5 of Manual S, the steps to selecting a heat pump are shown in Figure 28.

Supplemental Heat – The amount of electric resistance heat beyond the heat pump capacity required to meet the peak heating load.

Emergency Heat – The amount of electric resistance heat required to meet the peak heating load should the heat pump fail.

Once the cooling equipment selection has been made, a balance-point diagram can be created from the manufacturer's published performance data and the heating season outdoor design temperatures. The balance-point diagram is used to determine the amount of supplemental and emergency heat required. Figure 29 shows the development of the balance-point diagram. The capacity line (A) is provided by the manufacturer in the detailed equipment specifications. The load line (B) is created by drawing a straight line from the design heating load at the outdoor design temperature to 0 Btu/h load at 65°F outdoor temperature. Emergency heat (C) is the total design heating load. Supplemental heat (D) is the difference between the capacity line and the design heating load.

4.1. Air Source Heat Pump Selection

The first step highlighted in Figure 30 instructs the designer to define the cooling design parameters from the cooling loads of the Manual J procedure. The cooling design information developed by Burdick (2011) from the Manual J procedures is shown in Figure 31.

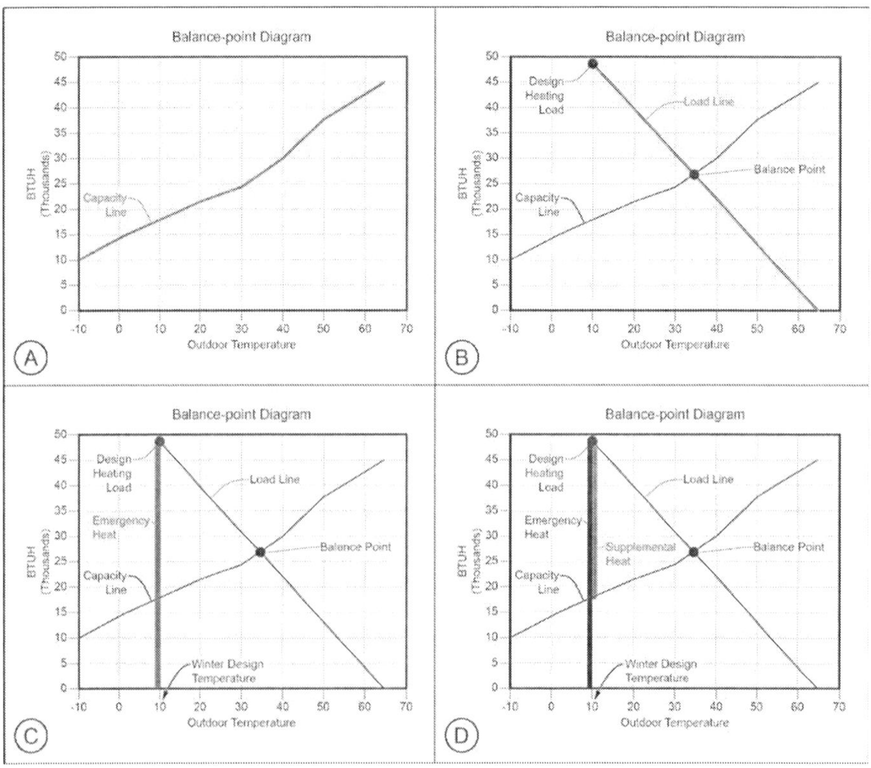

Figure 29. Development of balance-point diagram.

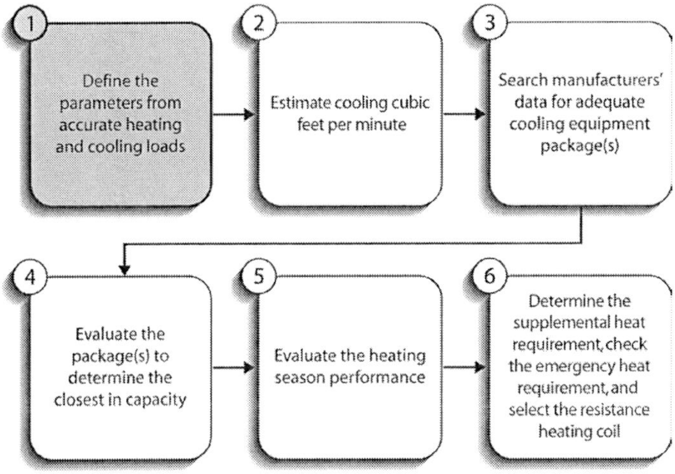

Figure 30. Air source heat pump selection step 1.

Manual J	Heating Load	23,640 Btu/h
	Sensible Cooling Load	16,630 Btu/h
	Latent Cooling Load	4,060 Btu/h
Outdoor Conditions	Cooling Dry-bulb Temperature	92 °F
	Heating Dry-bulb Temperature	38 °F
Indoor Conditions	Cooling Dry-bulb Temperature	75 °F
	Cooling Wet bulb Temperature	63 °F
	Heating Dry-bulb Temperature	70 °F
Airflow Estimates	Cubic Feet per Minute	719 cfm

Figure 31. Orlando House design information.

4.2. Heat Pump Cooling Capacity Selection

Step 2 of the selection process, highlighted in Figure 32, instructs the designer to estimate the cooling cfm. The methods outlined in Section 1-3 of Manual S are used to estimate the initial cfm airflow.

Equation 7 applies the cooling design parameters from Figure 31 to Equation 1, resulting in an SHR of 0.80.

$$\text{SHR} = \frac{\text{Sensible Load}}{\text{Total Load}} = \frac{16.63\frac{\text{MBtu}}{\text{h}}}{20.69\frac{\text{MBtu}}{\text{h}}} = 0.80 \tag{7}$$

From the TD value table (Figure 11), the SHR value of 0.80 is the break point between the TD values of 19 and 21. Note that a higher TD value will lead to a lower target cfm value, and a lower cfm value is desirable for a humid climate. Therefore, a TD of 21 was selected for this Orlando House example in Florida. This is the type of design consideration that can be applied to the equipment selection procedure.

With the TD value for SHR = 0.80 of 21 selected, these values can now be applied to the sensible heat equation (Equation 2) as shown in Equation 8.

$$\text{cfm Target} = \frac{\text{Manual J Sensible Load}}{1.1 \times \text{TD}} = \frac{16{,}630}{1.1 \times 21} = 719 \tag{8}$$

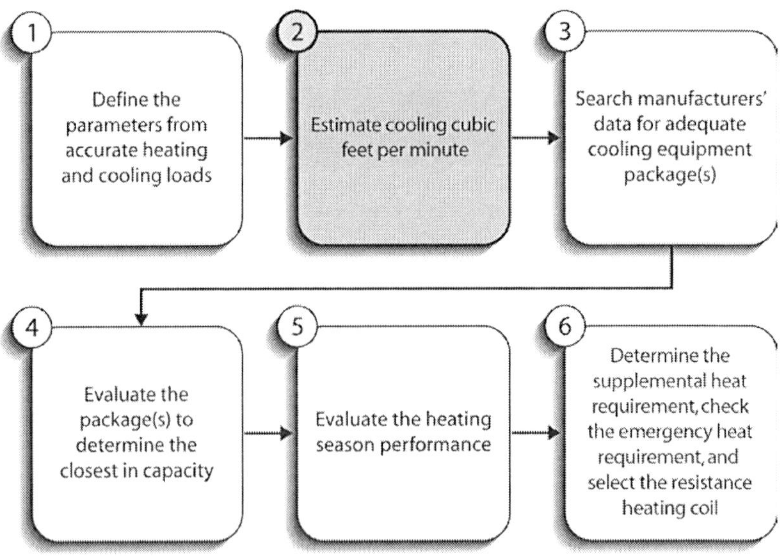

Figure 32. Air source heat pump selection step 2.

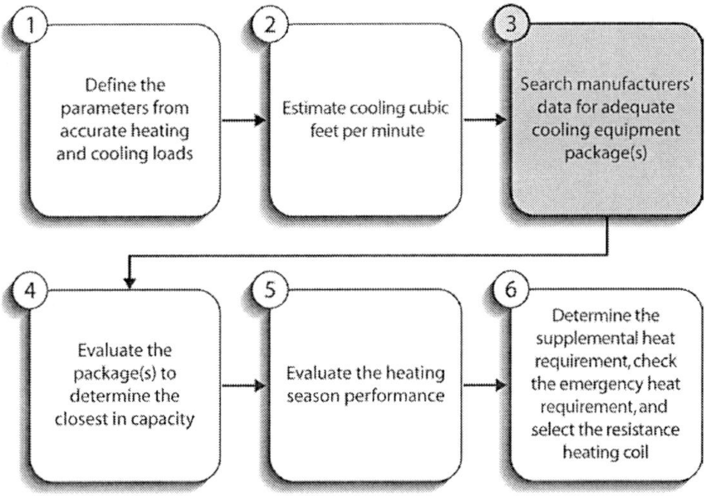

Figure 33. Air source heat pump selection step 3.

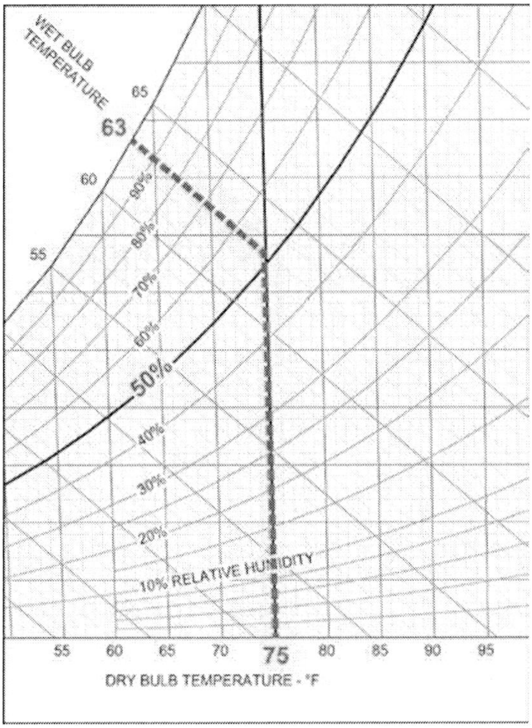

Figure 34. Wet-bulb temperature on sea-level psychometric chart.

Step 3, highlighted in Figure 33, is to search the manufacturer's data for an adequate cooling equipment package. To search the fictitious example of a cooling equipment package detailed cooling capacities, a nominal size must be determined. Considering that 12.00 MBtu/h is equivalent to 1 nominal ton of cooling, the 20,690 Btu/h total cooling load for the Orlando House leads to a 2-ton nominal cooling system as the target size to begin the equipment search.

The EWB temperature for the indoor coil can be determined on the psychometric chart from the Manual J conditions of 75°F indoor design dry-bulb temperature and 50% relative humidity, as shown in Figure 34.

The detailed cooling capacities tables give the data required for step 4 (see Figure 35). The outside design temperature for the Orlando House used in the Manual J load calculation was 92°F. The 92°F outdoor design temperature will require interpolation of the manufacturer's published data between the 85°F and 95°F condenser entering air temperature points.

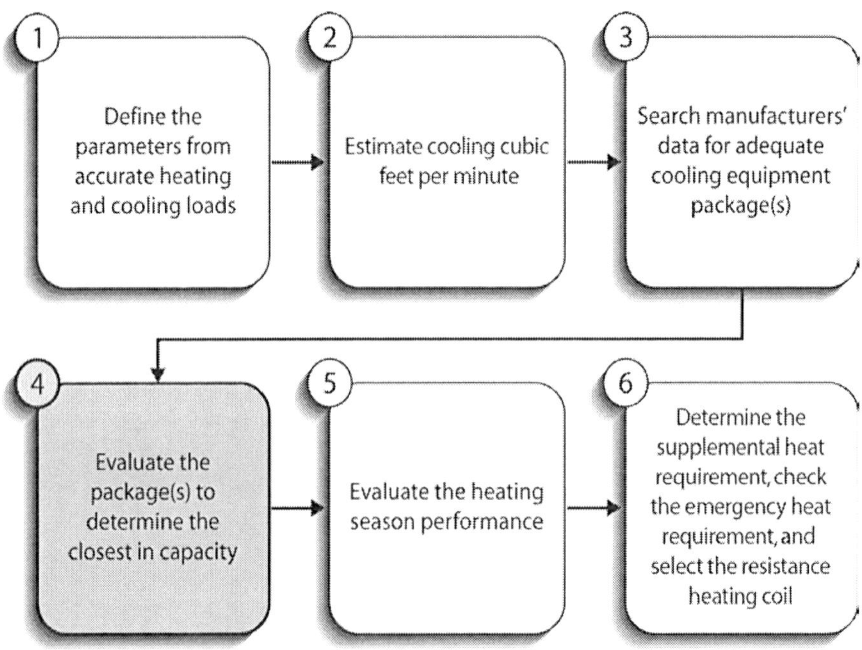

Figure 35. Cooling equipment selection step 4.

The fictitious example heat pump detailed cooling capacities table (Figure 36) is searched at the closest published cfm data to the initial cfm airflow estimate of 719 cfm. Figure 36 highlights the 700 cfm and 63°F entering evaporator wet-bulb temperature row. The published 700 cfm data in Figure 36 for both total and sensible capacities appear to be within the range of the load requirements. Figure 37 shows the linear interpolation of the values to determine the capacity at 92°F. This analysis shows that the capacity of this package at 92°F will meet the total, sensible, and latent loads at 700 cfm. Utilizing a spreadsheet to handle the multiple interpolation calculations makes the procedure more manageable, as shown in Figure 37.

DETAILED COOLING CAPACITIES# — **24XYZ**

EVAPORATOR AIR		CONDENSER ENTERING AIR TEMPERATURES °F (°C) 75 (23.9)			85 (29.4)			95 (35)			105 (40.6)		
CFM	EWB °F (°C)	Capacity MBtuh Total	Capacity MBtuh Sens‡	Total Sys. KW**	Capacity MBtuh Total	Capacity MBtuh Sens‡	Total Sys. KW**	Capacity MBtuh Total	Capacity MBtuh Sens‡	Total Sys. KW**	Capacity MBtuh Total	Capacity MBtuh Sens‡	Total Sys. KW**
24XYZ01234 Outdoor Section With AB3CDE456 Indoor Section													
600	72 (22.2)	27.18	14.15	1.44	25.86	13.53	1.63	24.48	12.96	1.82	23.07	12.47	2.06
	67 (19.4)	24.67	16.09	1.42	23.49	16.34	1.60	22.28	15.78	1.80	20.95	15.14	2.02
	63 (17.2)††	22.75	18.19	1.41	21.86	15.83	1.59	20.80	15.38	1.78	19.28	14.42	2.00
	62 (16.7)	22.41	19.69	1.40	21.34	19.52	1.58	20.31	18.40	1.78	19.02	17.76	2.00
	57 (13.9)	21.64	21.64	1.39	20.22	20.22	1.57	19.33	19.33	1.77	18.39	18.39	1.99
700	72 (22.2)	28.13	15.04	1.47	28.75	14.46	1.65	26.30	13.96	1.85	28.79	13.54	2.07
	67 (19.4)	25.58	18.21	1.45	24.32	17.62	1.63	22.90	17.36	1.83	21.61	16.26	2.05
	63 (17.2)††	23.61	17.41	1.43	**22.44**	**16.82**	1.61	**21.20**	**16.20**	1.81	19.91	15.56	2.03
	62 (16.7)	23.27	21.24	1.43	22.13	20.72	1.61	20.94	20.36	1.80	19.70	19.32	2.02
	57 (13.9)	22.33	22.33	1.42	21.43	21.43	1.60	20.46	20.46	1.80	19.43	19.43	2.02
735	72 (22.2)	29.43	15.32	1.48	27.81	14.74	1.66	25.54	14.13	1.85	24.09	13.59	2.08
	67 (19.4)	25.88	19.65	1.48	24.57	18.05	1.64	23.22	17.42	1.83	21.81	16.77	2.06
	63 (17.2)††	23.86	17.82	1.44	22.87	17.22	1.62	21.41	16.56	1.82	20.10	15.93	2.03
	62 (16.7)	23.53	21.53	1.44	22.38	21.26	1.62	21.17	20.56	1.81	19.93	19.51	2.03
	57 (13.9)	22.74	22.74	1.43	21.69	21.60	1.61	20.81	20.51	1.81	19.78	19.78	2.03
800	72 (22.2)	29.81	15.63	1.48	27.45	15.23	1.67	20.93	14.51	1.87	24.35	13.97	2.09
	67 (19.4)	25.35	19.45	1.47	24.88	18.63	1.65	23.59	18.19	1.85	22.13	17.52	2.07
	63 (17.2)††	24.28	18.58	1.45	23.08	17.94	1.63	21.75	17.30	1.83	20.45	16.63	2.06
	62 (16.7)	23.98	22.97	1.45	22.80	22.27	1.63	21.56	21.51	1.83	20.35	20.35	2.05
	57 (13.9)	23.44	23.44	1.44	22.46	22.46	1.63	21.42	21.42	1.83	20.31	20.31	2.05

Figure 36. Fictitious example: detailed heat pump cooling capacities table, 700 cfm.

City: Orlando, FL	Cooling Outdoor Design Temperature = 92 °F										
	Total Cooling Design Load = 20.7 MBtu/h										
	Sensible Cooling Design Load = 17.4 MBtu/h										
	Condenser Entering Air Temperatures °F										
	85	86	87	88	89	90	91	92	93	94	95
Total Cooling (CFM)	Total Capacity (MBtu/h)										
700	22.44	22.32	22.19	22.07	21.82	21.82	21.70	**21.57**	21.45	21.32	21.20
735	22.65	22.53	22.40	22.28	22.15	22.02	21.90	21.77	21.64	21.52	21.39
800	23.05	22.92	22.79	22.66	22.53	22.40	22.27	22.14	22.01	21.88	21.80
	Sensible Capacity (MBtu/h)										
700	16.82	16.76	16.70	16.63	16.57	16.51	16.45	**16.39**	16.32	16.26	16.20

Figure 37. Interpolation table for cooling capacities at 92°F outdoor design temperature.

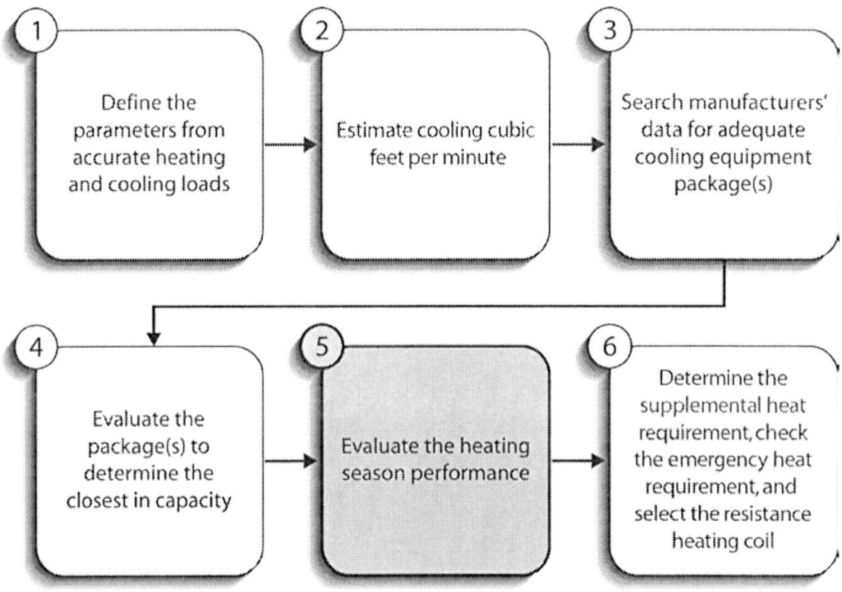

Figure 38. Cooling equipment selection step 5.

HEAT PUMP HEATING PERFORMANCE

24XYZ

INDOOR AIR		OUTDOOR COIL ENTERING AIR TEMPERATURES °F (°C)																	
		-3 (-19.4)			7 (-13.9)			17 (-8.3)			27 (-2.8)			37 (2.8)			47 (8.3)		
EDB °F (°C)	CFM	Capacity MBtuh Total	Capacity MBtuh Integ*	Total Sys. KW†	Capacity MBtuh Total	Capacity MBtuh Integ*	Total Sys. KW†	Capacity MBtuh Total	Capacity MBtuh Integ*	Total Sys. KW†	Capacity MBtuh Total	Capacity MBtuh Integ*	Total Sys. KW†	Capacity MBtuh Total	Capacity MBtuh Integ*	Total Sys. KW†	Capacity MBtuh Total	Capacity MBtuh Integ*	Total System KW†
24XYZ01234 Outdoor Section With ABCDE456 Indoor Section																			
65 (18.3)	600	9.80	9.06	1.48	12.27	11.28	1.58	15.08	13.75	1.70	18.30	16.20	1.83	21.96	19.98	1.93	26.11	26.11	2.13
	700	9.98	9.18	1.47	12.45	11.44	1.56	15.31	13.96	1.67	18.59	16.51	1.79	22.31	20.30	1.95	26.46	26.46	2.15
	735	10.01	9.21	1.47	12.50	11.49	1.56	15.38	14.02	1.66	18.68	16.59	1.78	22.40	20.38	1.94	26.56	26.56	2.15
	800	10.08	9.28	1.46	12.59	11.57	1.56	15.49	14.12	1.65	18.80	16.70	1.77	22.54	20.51	1.93	26.67	26.67	2.14
70 (21.1)	600	9.62	8.85	1.56	12.01	11.04	1.67	14.77	13.47	1.76	17.93	15.93	1.92	21.54	19.60	2.08	25.64	25.64	2.29
	700	9.77	8.99	1.56	12.19	11.20	1.66	15.00	13.67	1.75	18.22	16.18	1.88	21.88	19.92	2.03	26.00	26.00	2.24
	735	9.81	9.03	1.55	12.24	11.25	1.64	15.06	13.73	1.70	18.31	16.26	1.87	21.98	20.00	2.02	26.09	26.09	2.23
	800	9.88	9.09	1.55	12.33	11.33	1.64	15.18	13.84	1.74	18.44	16.38	1.86	22.13	20.14	2.01	26.22	26.22	2.23
75 (23.9)	600	9.20	8.47	1.64	11.75	10.80	1.75	14.45	13.18	1.87	17.57	15.60	2.01	21.12	19.22	2.17	25.15	25.15	2.38
	700	9.51	8.75	1.63	11.92	10.96	1.73	14.68	13.38	1.84	17.85	15.86	1.97	21.40	19.53	2.12	25.52	25.52	2.33
	735	9.55	8.79	1.63	11.98	11.01	1.73	14.75	13.45	1.83	17.93	15.93	1.96	21.57	19.63	2.11	25.62	25.62	2.32
	800	9.63	8.86	1.63	12.07	11.09	1.72	14.86	13.55	1.82	18.07	16.05	1.94	21.72	19.76	2.10	25.75	25.75	2.31

Figure 39. Heat pump heating performance table at 700 cfm.

	Heating Outdoor Design Temperature = 38 °F										
	Heating Design Load = 23.6 MBtu/h										
	Condenser Entering Air Temperatures °F										
	37	38	39	40	41	42	43	44	45	46	47
Total Heating (CFM)	Total Capacity (MBtu/h)										
700	22.31	22.73	23.14	23.56	23.97	24.39	24.80	25.22	25.63	26.05	26.46
735	22.39	22.80	23.22	23.63	24.05	24.46	24.88	25.29	25.70	26.12	26.53
800	22.54	22.95	23.37	23.78	24.19	24.61	25.02	25.43	25.84	26.26	26.70

Figure 40. Interpolation table for heating capacities at 38°F outdoor design temperature.

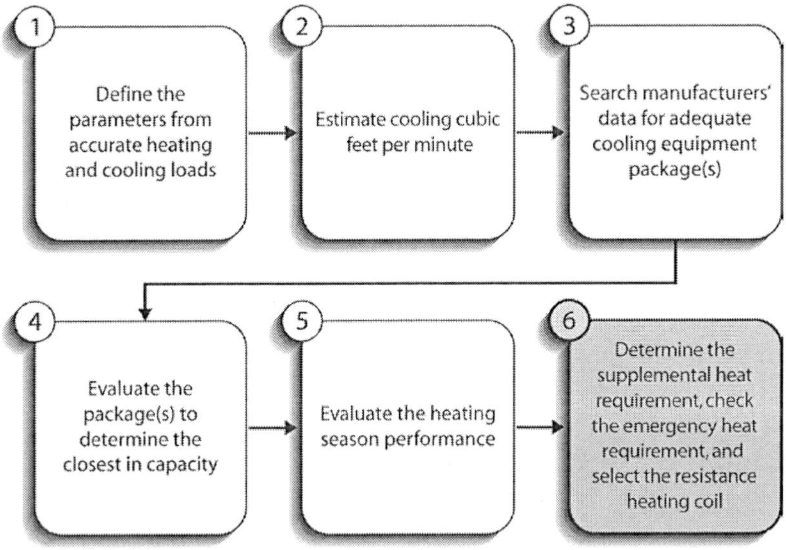

Figure 41. Cooling equipment selection step 6.

4.3. Heat Pump Heating Capacity Selection

To complete step 5 highlighted in Figure 38 and to evaluate the heating season performance of the heat pump, the heating capacity must be determined from the heat pump heating performance table as shown in Figure 39, and interpolation to the 38°F heating design temperature will be required. The indoor heating design temperature from Manual J is 70°F, and the design cfm determined in the cooling selection is 700 cfm.

Interpolation of the capacity at 38°F is shown in Figure 40. This equipment's capacity of 22.73 MBtu/h closely matches the heating load of 23.64 MBtu/h, and the balance will be made up with supplemental heat as determined in step 6 (Figure 41). Step 6 is to determine the supplemental and emergency heat requirements from the balance-point diagram.

The balance-point diagram, Figure 42, for the selected outdoor heat exchanger and fan coil shows a small 3,000 Btu/h supplemental heat requirement.

Supplemental and emergency heat are supplied by electric resistance heat and are sized in kilowatts, requiring a conversion from Btu/h. The 23.64 MBtu/h total load requirement converts to 6.9 kW emergency heat, and the supplemental heat requirement is less than 1 kW. This supplemental heating capacity will be specified in the sizing of the fan coil.

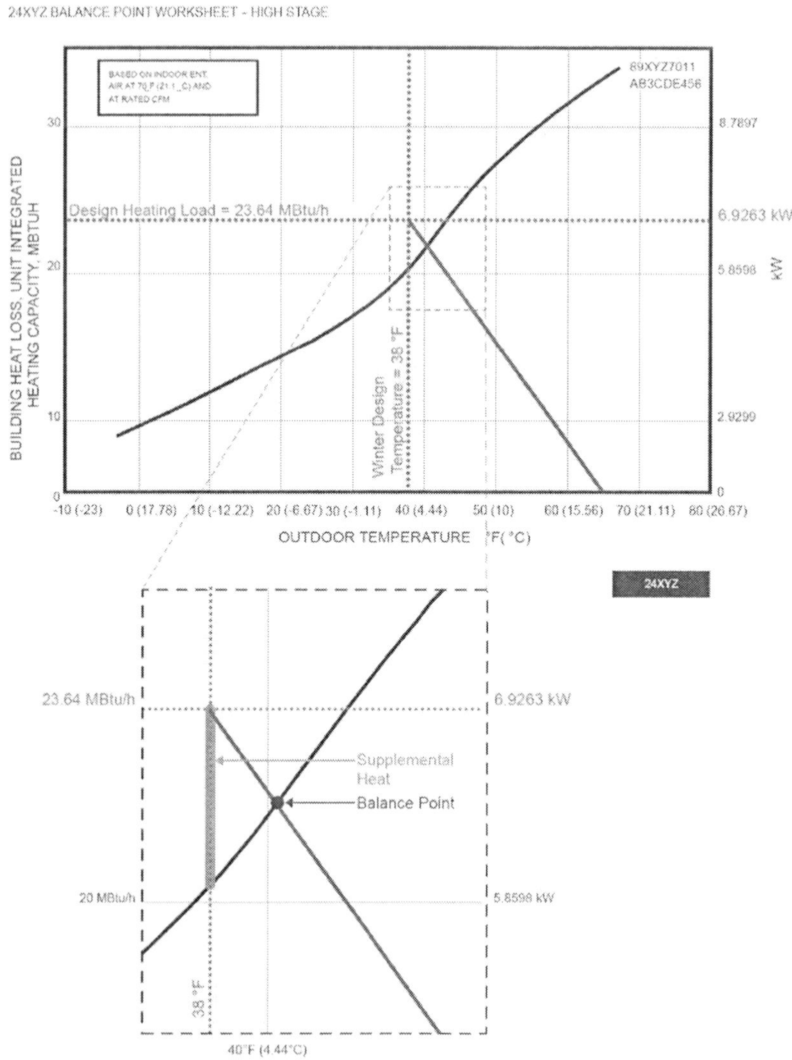

Figure 42. Balance-point diagram for the Orlando House.

The electric resistance coils will be specified with the fan coil sizing shown in Figure 43. The available coils are 5, 8, and 10 kW. The 8 kW is larger than the required 6.9 kW, but this is resolved by staged resistance heat providing resistance heat in eight individual 1-kW increments.

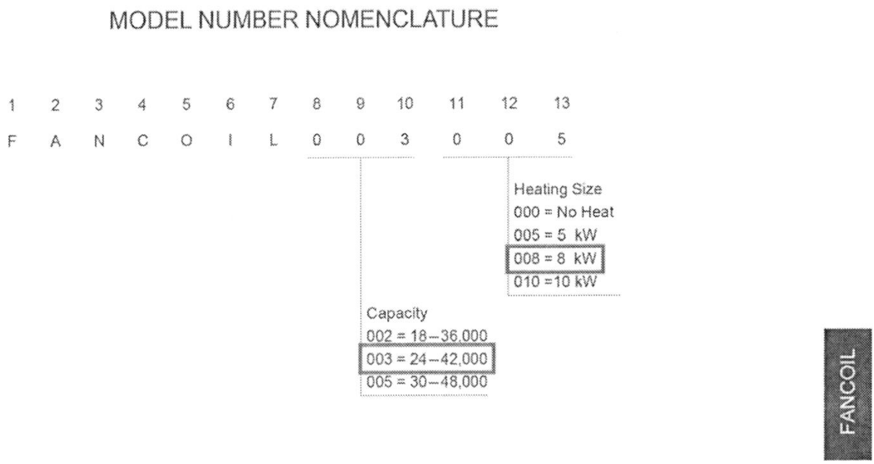

Figure 43. Fan coil nomenclature.

We now have a heat pump and fan coil package that will meet the heating and cooling load needs of the Orlando House. The next step in right-sizing the HVAC system is to use the blower performance data of the fan coil to create the air delivery duct design.

CONCLUSION

Right-sizing of an HVAC system is the selection of equipment and the design of the air distribution system to meet the accurate predicted heating and cooling loads of the house. The equipment effective capacity should match the loads generated in the load calculation procedure within the sizing guidelines. Selection of the equipment can have a substantial impact on the overall efficiency and operating costs of the installed system.

Note that the design of a space conditioning system is an iterative process, and once a duct system configuration and preliminary design are completed, different equipment may need to be selected to achieve the proper airflow to satisfy the room-by-room heating and cooling loads. The design loads along with the indoor and outdoor conditions information needed for equipment selection are all taken from the inputs used in the Manual J procedure. Faithful adherence to the conditions used in Manual J is required for accurate equipment selection.

The selection of heating and cooling equipment is dependent on accurate heating and cooling load calculations. From this point onward, in the design process, the actual performance data of the equipment selections and room-by-room loads from the Manual J calculations should be used to design the duct system.

References

[1] Burdick, A. (2011). *Strategy Guideline: Accurate Heating and Cooling Load Calculations*. Pittsburgh: IBACOS. Prepared for the U.S. Department of Energy Building America Program. www.nrel.gov/docs/fy11osti/51603.pdf.

[2] IECC (2009). International Energy Conservation Code. Washington, DC: International Code Council.

[3] Rutkowski, H. (1995). *Manual S—Residential Equipment Selection*. Arlington, VA: Air Conditioning Contractors of America.

[4] Rutkowski, H. (2006). *Manual J—Residential Load Calculation*, 8th edition, Version 2. Arlington, VA: Air Conditioning Contractors of America.

[5] Rutkowski, H. (2009a). *Manual D—Residential Duct Systems*, 3rd edition, Version 1.00. Arlington, VA: Air Conditioning Contractors of America.

[6] Rutkowski, H. (2009b). *Manual T—Air Distribution Basics for Residential and Small Commercial Buildings*. Arlington, VA: Air Conditioning Contractors of America.

INDEX

Q

R

S

T